Contents

AF874506

About this Publication

This publication provides the Theoretical Knowledge (TK) required in Meteorology for non-commercial flight operations under Visual Flight Rules (VFR). It is based on the 2015 PPL and LAPL Theoretical Knowledge (TK) Meteorology syllabus published by the UK CAA and accepted by EASA and it is considered suitable for use in conjunction with training courses for the following pilot licences:

- EASA Private Pilot Licence (PPL) (Aeroplane)
- EASA Light Aircraft Pilot Licence (LAPL) (Aeroplane)
- CAA Private Pilot Licence (PPL) (Aeroplane)
- CAA Light Aircraft Pilot Licence (LAPL) (Aeroplane)
- UK National PPL (NPPL) (Aeroplane)
- ICAO-compliant PPL (Aeroplane) or equivalent licence
- Core knowledge for EASA and CAA Commercial Pilot Licence (CPL) (Aeroplane)
- Foundation knowledge for EASA and CAA Airline Transport Pilot Licence (ATPL) (Aeroplane)
- Foundation knowledge for EASA and CAA Multi-crew Pilot Licence (MPL) (Aeroplane)

This publication also provides foundation knowledge for an Air Transport Operations/ Management degree or similar academic qualification.

About the Author

Jeremy M Pratt took his first flying lesson at the age of 14, paid for by working in the hangar and radio unit at his local airfield. He gained his pilot's licence shortly after his 18th birthday after being awarded an Esso/Air League Flying Scholarship, became a flying instructor at 19 and a commercial pilot at the age of 20.

Since then he has taught (and continues to teach) pilots for a wide range of licences – both private and professional – as well as associated ratings and qualifications including night, instrument, tailwheel and multi-engine flying and he has flown General Aviation aircraft professionally in a number of other roles including pleasure flights, traffic reporting, aerial photography and aerial survey. He has owned and co-owned a number of General Aviation aircraft and flown a variety of aircraft types from Tiger Moth biplane to Cessna Citation jet, as well as trying-out helicopter, microlight and balloon flying. He also enjoys the highs (and occasional lows) of flying gliders.

The author's first flying training books were published in 1992, since when they have sold several hundred thousand copies world-wide and have been translated into a number of languages. He has also authored and co-authored around 25 additional aviation training books as well as contributing to various aviation publications including Flight Training News (FTN), for which he flight tests various aircraft types.

The author works with various aviation authorities and organisations on training and safety issues. He was also part of the team that produced the 2015 CAA/EASA PPL and LAPL syllabi and he sits on the CAA's PPL Theoretical Knowledge Working Group. He flies whenever he can find the time – for instructing, for business, for pleasure and for the sheer joy of flight.

Image Acknowledgements

Rights are reserved for all images used in this book. Where known the appropriate rights or copyright holder is listed below. AFE Ltd = Airplan Flight Equipment Ltd; JMP = Jeremy M Pratt. In a small number of cases we have been unable to trace the rights holder for an image, in which case it is marked 'ukn' (unknown).

M1 The Atmosphere

M1.1 ©AFE Ltd; M1.1a ©NASA; M1.2 ©AFE Ltd; M1.3 ©JMP; M1.4 ©AFE Ltd

M2 Temperature, Pressure and Density

M2.1 ©AFE Ltd; M2.2 ©AFE Ltd; M2.3 ©schusterbauer.com; M2.4 ©AFE Ltd; M2.5 ©AFE Ltd; M2.6 ©AFE Ltd; M2.7 ©AFE Ltd; M2.8 ©AFE Ltd; M2.9 ©AFE Ltd/JMP; M2.10 ©AFE Ltd/JMP; M2.11 ©AFE Ltd / JMP; M2.12 ©AFE Ltd;

M3 Altimetry

M3.0 UKN; M3.1 ©AFE Ltd; M3.2 ©AFE Ltd; M3.3 ©AFE Ltd; M3.4 ©AFE Ltd; M3.5 ©AFE Ltd; M3.6 ©AFE Ltd;

M4 Altimetry

M4.1 ©AFE Ltd; M4.2 ©AFE Ltd; M4.3 ©AFE Ltd, M4.4 ©AFE Ltd; M4.5 ©AFE Ltd; M4.6 ©AFE Ltd; M4.7 ©AFE Ltd; M4.8 ©AFE Ltd; M4.9 ©AFE Ltd; M4.10 n/a; M4.11 ©AFE Ltd; M4.12 ©AFE Ltd/JMP; M4.13 ©AFE Ltd/JMP; M4.14 ©AFE Ltd/JMP; M4.15 ©AFE Ltd; M4.16 ©AFE Ltd; M4.17 ©AFE Ltd; M4.18 ©AFE Ltd; M4.19 ©AFE Ltd

M5 Clouds and Precipitation

M5.1 ©AFE Ltd; M5.2 ©AFE Ltd; M5.3 ©AFE Ltd; M5.4 ©JMP; M5.5 ©JMP; M5.6 ©AFE Ltd/JMP; M5.7 ©pR13S7; M5.8 ©Ethan Daniels; M5.9 ©AFE Ltd; M5.10 ©JMP; M5.11 ©Happy Dragon; M5.12 ©JMP; M5.13 ©JMP; M5.14 ©Dariusz Leszczynski; M5.15 ©JMP; M5.16 ©JMP; M5.17 ©JMP; M5.18 ©JMP; M5.19 ©vololibero; M5.20 ©Bjoern Wylezich; M5.21 ©JMP; M5.22 ©JMP; M5.23 ©JMP; M5.24 ©JMP; M5.25 ©JMP; M5.26 ©JMP; M5.27 ©JMP; M5.28 ©AFE Ltd; M5.28a ©JMP; M5.29 ©Thomas Bethge; M5.30 ©AFE Ltd; M5.31 ©JMP

M6 Visibility

M6.1 ©AFE Ltd; M6.2 ©Roman Kosolapov; M6.3 ©Lijuan Guo; M6.4 ©AFE Ltd; M6.5 ©JMP; M6.6 ©NORRIE3699; M6.7 ©Met Office; M6.8 ©JMP; M6.9 ©AFE Ltd; M6.10 ©JMP; M6.11 ©JMP; M6.12 ©JMP; M6.13 ©AFE Ltd/JMP; M6.14 ©AFE Ltd; M6.15 ©Jordan Bridge; M6.16 ©JMP; M6.17 ©JMP; M6.18 ©JMP; M6.19 ©AFE Ltd/Jordan Bridge; M6.20 ©AFE Ltd/JMP; M6.21 ©AFE Ltd/JMP; M6.22 ©AFE Ltd; M6.23 ©Alexey Lesik; M6.24 ©AFE Ltd;

M7 Air Masses

M7.1 ©AFE Ltd; M7.2 ©AFE Ltd; M7.3 ©JMP

M8 Low Pressure Systems

M8.1 ©AFE Ltd; M8.2 ©AFE Ltd; M8.3a ©AFE Ltd; M8.3b ©AFE Ltd; M8.3c ©AFE Ltd; M8.3d ©AFE Ltd; M8.4 ©AFE Ltd; M8.5 ©AFE Ltd; M8.6 ©AFE Ltd; M8.7 ©AFE Ltd; M8.8 ©AFE Ltd; M8.9 ©AFE Ltd; M8.10 ©AFE Ltd; M8.11 ©AFE Ltd; M8.12 ©AFE Ltd; M8.13 ©AFE Ltd; M8.14 ©AFE Ltd; M8.15 ©Met Office

M9 High Pressure Systems

M9.1 ©AFE Ltd; M9.2 ©AFE Ltd; M9.3 ©JMP; M9.4 ©AFE Ltd; M9.5 ©JMP; M9.7 ©JMP; M9.8 ©AFE Ltd; M9.9 ©AFE Ltd

M10 Hazardous Weather Conditions: icing

M10.1 ©AFE Ltd; M10.2 UKN; M10.3 ©AFE Ltd; M10.4a ©JMP; M10.4b ©JMP; M10.6 courtesy Met Office; M10.7 ©JMP; M10.8 ©JMP; M10.9 ©AFE Ltd; M10.10 ©CASA; M10.11 ©CAA / EASA

M11 Hazardous Weather Conditions: Thunderstorms

M11.1 ©Joanne Treacy; M11.2 ©JMP; M11.3 ©JMP; M11.4 ©AFE Ltd; M11.5 ©AFE Ltd; M11.6 ©JMP; M11.7 ©AFE Ltd; M11.8 ©AFE Ltd; M11.9 ©Met Office; M11.9a ©Met Office; M11.10 ©JMP; M11.11 ©JMP; M11.12 ©JMP; M11.13 ©AFE Ltd; M11.14 ©JMP; M11.15 ©JMP; M11.16 ©JMP; M11.17 ©JMP; M11.18 ©AFE Ltd; M11.19 ©JMP; M11.20 ©JMP; M11.21 courtesy Met Office / ©AFE Ltd; M11.22 ©JMP; M11.23 ©JMP; M11.24a UKN; M11.25 ©JMP; M11.26 ©AFE Ltd/JMP; M11.27 ©JMP; M11.28 UKN; M11.29 courtesy Met Office; M11.30 ©James McBride

M12 Other Hazardous Weather Conditions

M12.1 ©AFE Ltd/JMP; M12.2 ©JMP; M12.3 ©JMP; M12.4 ©JMP; M12.5 ©AFE Ltd/JMP; M12.6 ©AFE Ltd/JMP; M12.7 ©AFE Ltd; M12.8 ©JMP; M12.9 ©AFE Ltd/JMP; M12.10 ©AFE Ltd/JMP; M12.11a ©JMP; M12.11b ©JMP; M12.11c ©JMP; M12.12 ©AFE Ltd/JMP; M12.13 ©AFE Ltd/JMP; M12.14 ©AFE Ltd; M12.15 ©AFE Ltd/JMP; M12.16 ©AFE Ltd/JMPM; M12.17 ©AFE Ltd/JMP; M12.18 ©Met Office; M12.19 ©Met Office; M12.20 courtesy Met Office; M12.21 ©AFE Ltd; M12.22 ©AFE Ltd; M12.23 ©JMP

M13 Meteorological Information

M13.1 ©Met Office; M13.2 ©AFE Ltd; M13.3 ©Met Office; M13.4 ©Met Office; M13.5 ©Met Office; M13.6 ukn; M13.7 courtesy Met Eireann; M13.8 ukn; M13.8a ©AFE Ltd; M13.8b ukn; M13.18c ©Jersey Met; M13.8d courtesy WAFC; M13.9 ©AFE Ltd; M13.9a ©AFE Ltd; M13.10 ©AFE Ltd; M13.11 ©AFE Ltd; M13.12 courtesy DGAC; M13.13 ©SkyDemon; M13.14 ukn; M13.15 ©Goodwood Aerodrome; M13.15a ©Charles Church Spitfires Ltd; M13.15b ©Met Office; M13.15c ukn; M13.15d ukn; M13.15e courtesy Met Office; M13.16 courtesy IAA; M13.17 ©AFE Ltd; M13.18 courtesy AIM Belgium; M13.19 ©JMP

M13 national Procedures

M14.1 ©AFE Ltd; M14.2 ©Met Office; M14.3 ©Met Office; M14.4 ©Met Office; M14.4 courtesy Met Office; M14.5 ©Met Office; M14.6 ©Met Office; M14.7 courtesy Met Office; M14.8 ©AFE Ltd; M14.9 courtesy Met Office; M14.10 ©AFE Ltd;

Appendix 2 – Supplementary Study Material

App2.1 ©AFE Ltd; App2.2 ©AFE Ltd

Acknowledgements

Creating a book is, rather like flying itself, an activity that needs the help and assistance of a whole range of people to bring the dream into reality. I have been fortunate over the years to have met, flown and worked with a range of very talented people in both the aviation and publishing worlds and many of them have, in one way or another, contributed to this book. The following are just some of the people and organisations who have directly contributed or worked on this book, and I offer my heartfelt thanks to them. To any person or organisation I may have inadvertently missed out, I additionally offer my apologies for the oversight.

AIM Belgium

Airplan Flight Equipment (AFE)

Wendy Barratt

Jordan Bridge

Charles Church Spitfires Ltd

Civil Aviation Authority (CAA)

Civil Aviation Safety Authority (CASA)

CHIRP

European Aviation Safety Agency (EASA)

GASCo

Goodwood Aerodrome

Irish Aviation Authority (IAA)

James McBride

Jersey Met

Met Eireann

Met. Office

Rob Taylor

Sky Demon

Dave Unwin

And, of course, my long-suffering family who continue to tolerate my obsession for all things flying for reasons I've never fully understood.

The appropriate image and photograph permissions and copyrights are listed in the 'Image Acknowledgements' section

This book is dedicated to Aidan Ruff, with whom I shared aeroplanes and aerial adventures around the UK, Europe and the USA for more than 25 years.

Blue Skies my friend.

Introduction

As long as the human race as existed, the weather has played a leading role in our day-to-day lives. More than 2,000 years ago, Aristotle wrote one of the earliest surviving texts on weather – *Meteorologica* – which identified many meteorological phenomena. Early sea farers crossing the oceans understood the importance of the weather to their voyages; but even so it wasn't until the development of widespread telegraphic communications less than 200 years ago, that weather forecasts could be calculated and distributed.

Today, weather forecasting is a global business worth billions and utilising cutting edge technologies; for example the UK Met Office super computer systems are amongst the 25 most powerful in the world, collecting 215 billion weather observations from all across (and above) the planet every day, and capable of more than 14,000 trillion arithmetic operations per second.

At a human level, most of us have become used to carrying on our day-to-day lives without interference from all but the most extreme weather conditions. Even so, the weather – both what is actually occurring and what is forecast to occur – remains the predominate factor when planning and carrying-out a Visual Flight Rules (VFR) flight which by definition must take place in Visual Meteorological Conditions (VMC). In terms of Threat and Error Management (TEM), the weather is a prime example of a 'Threat' – an external factor outside the control of the pilot. Misunderstanding of weather reports and forecasts, failing to check weather information at all, or flying into unsuitable weather conditions, all represent the potential for a significant Error on the part of a pilot.

Meteorology cannot be learnt 'parrot fashion'; *an understanding* of the basic mechanisms at work is required. It is these basics that are covered in the first sections of this meteorology course and it pays to work through these sections carefully. A sound understanding of meteorological principles will make the rest of the course that much easier.

Meteorology theoretical knowledge **does not** aim to turn pilots into weather forecasters – instead this meteorological course should equip you to:

1: *interpret weather information* and so make safe planning decisions based on it;

2: *assess the actual weather* occurring, and make sound decisions based on what you see.

This course concentrates on the practical aspects of aviation meteorology which affect flight planning and in-flight decisions. To this end I have avoided, where possible, jargon and purely theoretical concepts. Furthermore, sample weather forecasts and weather reports are used extensively to illustrate weather phenomena. The intention is to give you early exposure to the type of weather information you will access as a pilot; the main features of each extract are described in the accompanying captions. If you want to fully decode a real-life forecast or report, the necessary information can be found in the 'Meteorological Information' section.

As communication technology has developed from the electric telegraph to the tablet and smart phone, there has been an exponential proliferation in weather information available to the pilot. Learning how to filter this information for the most relevant and reliable data is becoming an essential skill for any pilot. The sooner you start to 'self-brief' weather information before flight, and compare that information to what you encounter in-flight, the better equipped you will be to make good flight planning decisions and enjoy your flying.

Jeremy M Pratt
July 2024

Aeronautical Knowledge – Meteorology

Jeremy M Pratt

ISBN 9781800350670

Published by:
Airplan Flight Equipment Ltd

AFE
afeonline.com

This book is a guide to the CAA and EASA Meteorology Theoretical Knowledge syllabus for the Private Pilot Licence (PPL) and Light Aircraft Pilot Licence (LAPL) for aeroplanes.

This book is not intended to be an authoritative document and it does not in any way over-rule or counter instruction from an approved, declared or registered training organisation or information or guidance produced by the Civil Aviation Authority (CAA), European Aviation Safety Agency (EASA) or by any other Competent Authority. Full reference should be made to applicable UK and EU regulations, Implementing Regulations and Acceptable Means of Compliance and Guidance Material (AMC/GM); rules, procedures, limitations and guidance in documents produced by Competent Authorities, including but not limited to national law and guidance. Aeronautical Information Publications (AIP), Aeronautical Information Circulars (AIC) and the applicable Aircraft Flight Manual (AFM) – alternatively the Pilots Operating Handbook/Flight Manual (POH/FM) – must also be complied with at all times. Requirements of the aircraft operator, for example in an Operations Manual or flying order book, must also be complied with.

At the end of 2020 the UK formally left EASA; however for practical purposes EU and EASA regulations have been taken into UK law and there are few practical differences between UK and EASA aviation law. Where there are significant differences, these are noted in the 'National Procedures' section of this book.

Aeronautical information, including that relating to Meteorology, can – and does – change frequently and it is the pilot's responsibility to remain up-to-date with such changes. The provisions of good airmanship and safe operating practice should be adhered to at all times.

Whilst every care has been taken in compiling this publication, relying where possible on authoritative and official information sources, the publisher and editorial team will not be liable in any way for any errors or omission whatsoever.

First edition 2024

Aeronautical Knowledge –
Meteorology

ISBN 9781800350670

Printed in Bulgaria by Multiprint

Airplan Flight Equipment Ltd
1a Ringway Trading Estate,
Shadowmoss Road,
Manchester M22 5LH UK

www.afeonline.com

MIX
Paper | Supporting responsible forestry
FSC® C100431

M1

The Atmosphere

Composition of the Atmosphere

The Troposphere

Progress Check

M1

The Atmosphere

Composition of the Atmosphere

By definition, 'aircraft' operate by flying through the air of the atmosphere. The **atmosphere** is a mixture of gases held to the earth by gravity. The earth's atmosphere is commonly considered to extend from the surface to an altitude of 100km (around 330,000ft), beyond which you are into the realms of outer space and aircraft become spacecraft. Compared with the distance from the centre of the earth to the surface (around 6,371km), the atmosphere is an extraordinarily thin layer of air. Nevertheless, it is this film of gases which protects the surface from the vacuum of space and makes life on our planet possible.

The approximate proportions of the gases (by volume) in a dry atmosphere are:

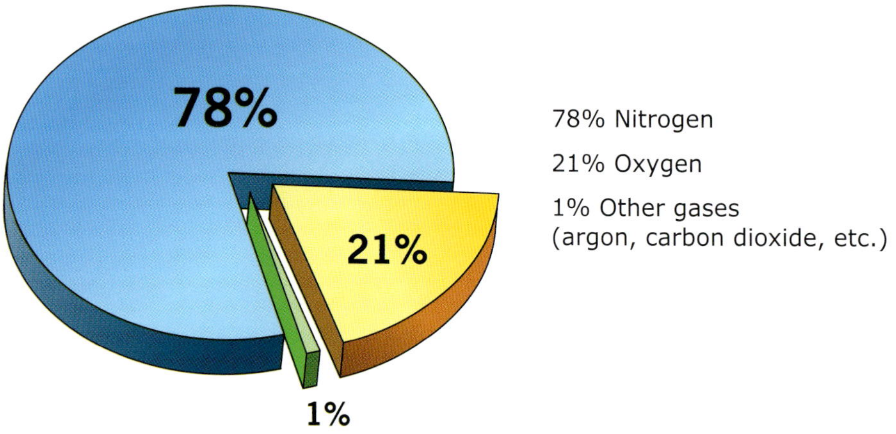

Figure M.1.1
The approximate proportions of the gases (by volume) in a dry atmosphere.

In fact, the atmosphere is never completely dry. The atmosphere contains fluctuating proportions of water vapour (which is an invisible gas), varying from close to zero in the cold, dry polar regions to a maximum of around 4% in the hot and humid parts of the world such as tropical seas. The vital importance of this water vapour in forming weather will be explored later.

Figure M1.1a
The earth's atmosphere, as seen from the International Space Station.

The atmosphere can be divided into layers. The layer closest to the earth and of most interest to the VFR pilot is the **troposphere**. The troposphere extends from the surface upwards to a boundary known as the **tropopause**. The tropopause marks the 'ceiling' of the troposphere and it is found at altitudes varying from around 20,000ft (6000m) over the polar regions to about 60,000ft (18,000m) over the equator. Over Europe, the average thickness of the troposphere is around 35,000ft (10,000m). Nevertheless the thickness of the troposphere, and so the level of the tropopause, varies daily and with the seasons.

Above the tropopause is the **stratosphere**. The tropopause effectively acts as a 'lid', keeping most of the water vapour – and so virtually all weather – in the troposphere. The stratosphere itself has almost no water vapour and extends from the tropopause to an average height of about 115,000ft above the surface.

Figure M1.2
The troposphere is approximately 20,000ft deep over the poles, 35,000ft deep over Europe and 60,000 deep over the tropics.

The Troposphere

It is this bottom layer of the atmosphere which contains virtually all the weather phenomena are which of interest to a VFR pilot. The troposphere contains around three quarters of the total mass of the atmosphere and approximately 99% of all water vapour in the atmosphere. Because of the presence of that water vapour, the troposphere is where almost all 'weather' – clouds, fog, mist, rain, snow etc. – is to be found, although occasionally a particular weather feature (such as the top of a thunderstorm) may push upwards through the tropopause.

The troposphere is thicker over the equator than over the poles, and in any given region it also extends higher in summer than in winter. Day-by-day, the level of the tropopause (the boundary between the troposphere and the stratosphere) rises and falls as high-pressure and low-pressure systems move across the earth.

One of the defining characteristics of the troposphere is that, on average, temperature in the troposphere reduces with increasing altitude. This relationship between altitude and temperature (increasing altitude leads to decreasing temperature) continues until the top of the troposphere is reached at the tropopause.

Figure M1.3
Almost all 'weather' is found in the troposphere, although a big thunderstorm may push through the tropopause.

Altitude	ISA Temperature
10,000ft	-5°C
9000ft	-3°C
8000ft	-1°C
7000ft	+1°C
6000ft	+3°C
5000ft	+5°C
4000ft	+7°C
3000ft	+9°C
2000ft	+11°C
1000ft	+13°C
Surface	+15°C

Figure M1.4
Temperature in the troposphere reduces with increasing altitude.

Progress Check

1. What is the average percentage of oxygen in a dry atmosphere?
2. What is the gas in the atmosphere which is responsible for most 'weather' – clouds, precipitation etc?
3. What is the name of the lowest layer of the atmosphere?
4. What is the name given to the boundary between the troposphere and and the stratosphere, and what is its average altitude over Europe?

These questions are intended to test knowledge and reinforce some of the key learning points from this section. In answering these questions, a 'pass rate' of around 80% should be the target.

Model answers are found at page M172.

M2 Temperature, Pressure and Density

M2 Temperature, Pressure and Density

Temperature variation in the atmosphere

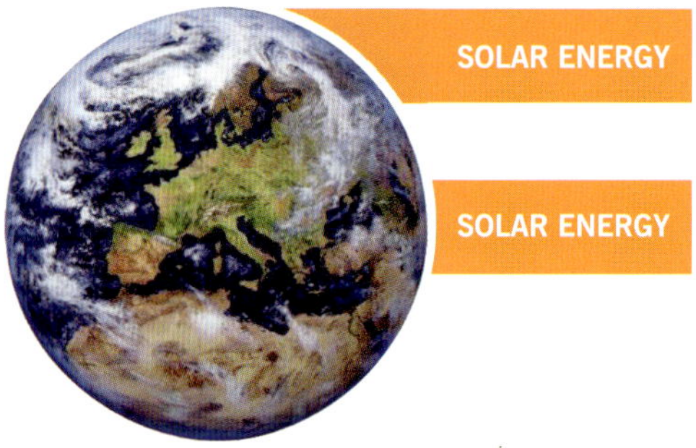

Figure M2.1
Solar energy is more concentrated at the surface of the equator than at the poles.

Pressure, temperature and water content are the key physical properties of the atmosphere, and much of the science (and art) of weather forecasting is based on measurement and prediction of these parameters. The interaction between these three properties largely determines the behaviour of the atmosphere and the type of weather likely to occur. It follows that a good knowledge of how pressure, temperature and water content interrelate is the key to understanding meteorology.

For practical purposes, the atmosphere itself acquires virtually no heat energy directly from the sun. Instead, radiation from the sun heats up the earth's surface, and some of this heat energy is transferred into the air which in contact with the surface.

For a number of reasons, including the virtually spherical shape of the earth, **solar energy** is more concentrated in tropical areas than in polar regions. So in the tropics the earth's surface gains a lot of heat energy from the sun and becomes very warm. Some of this heat is transferred to the atmosphere, and in effect the atmosphere is warmed from below. At the poles, the energy from the sun is spread over a much wider area and the surface remains cold. Hence the atmosphere in the vicinity of the poles is cold.

Figure M2.2
Because the earth is tilted on its axis, the Northern hemisphere experiences summer when it is winter in the southern hemisphere, and vice versa.

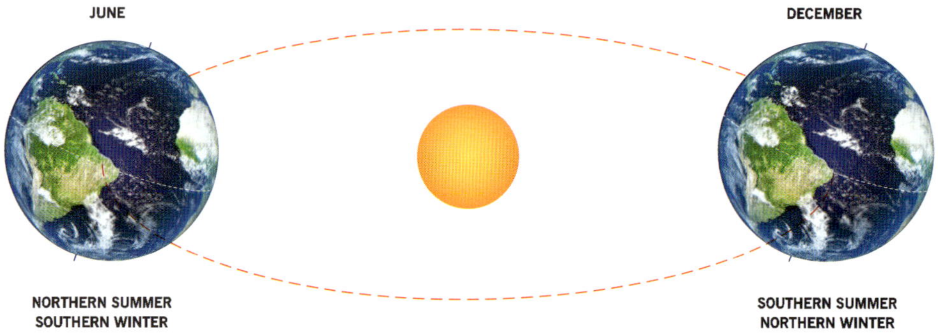

Because the earth is tilted on its axis (by about 23.5°) the northern and southern hemispheres receive differing concentrations of solar energy according to the position of the earth in its orbit around the sun (which causes the **seasons**). This is the reason why summer occurs in the northern hemisphere when it is winter in the southern hemisphere (and *vice versa*).

Different types of surface have different **heating properties**. Land surfaces tend to heat quickly and cool quickly, and a land surface is capable of changing from cold to hot and back in a matter of hours. Water changes temperature far more slowly, especially in the oceans which cover most of the earth's surface. The cycle of heating and cooling of the oceans can be measured in months, so there is often a relatively high sea temperature long after the summer (and therefore the strongest solar heating) has ended.

Figure M2.3
The seas and oceans heat up more slowly than the land, but retain heat energy for longer.

Conversely, having cooled slowly through the winter, it will be well into the late spring before sea temperatures begin to rise significantly.

As a general rule, within the troposphere (the lowest layer of the atmosphere) air temperature reduces as altitude increases. Even without getting into an aircraft, if you have ever hiked up a significant hill or mountain, you may well have noticed that it is noticeably colder at the top of the hill than at the bottom. The higher up the hill, the colder the air temperature. In mountainous regions, lying snow will last longest at the top of the mountains (and if the mountains are high enough, they may be covered by snow all year round).

In the troposphere, the average rate of **air temperature** change with changing altitude is about 2°C for every 1000ft of altitude change. So if the surface air temperature is +15°C, the temperature at an altitude of 1000ft on an average day would be +13°C. The rate of temperature change through the atmosphere is known as the **lapse rate,** and meteorologists take great interest in graphs showing the change of air temperature through the atmosphere at a given location and at specific time. The reduction in air temperature continues right up to the tropopause, above which the air temperature is assumed to remain constant regardless of altitude (at least at levels that aircraft can fly at). Therefore, there is no lapse rate in the tropopause and lower stratosphere.

Although the average lapse rate is around 2°C/1000ft, in the real world the actual lapse rate changes all the time and is likely to be different at different levels in the troposphere. The consequences of differing lapse rates are explored later.

For the scientifically-minded, a vertical column of air through which there is no change in temperature is called an 'isothermal' layer.

Still in the real world, air temperature does not always decrease with increasing altitude; sometimes it can do precisely the opposite. This situation, when air temperature actually increases with increasing altitude, is known as a **temperature inversion** (the normal lapse rate is inverted). A temperature inversion is likely to have a significant effect on meteorological conditions, and again is described in more detail later.

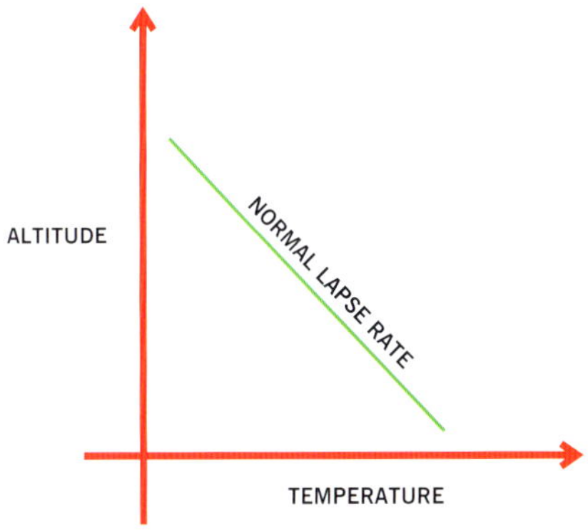

Figure M2.4
In the troposphere, as altitude increases, temperature normally decreases.

Figure M2.5
A temperature inversion, when temperature actually increases with increasing altitude.

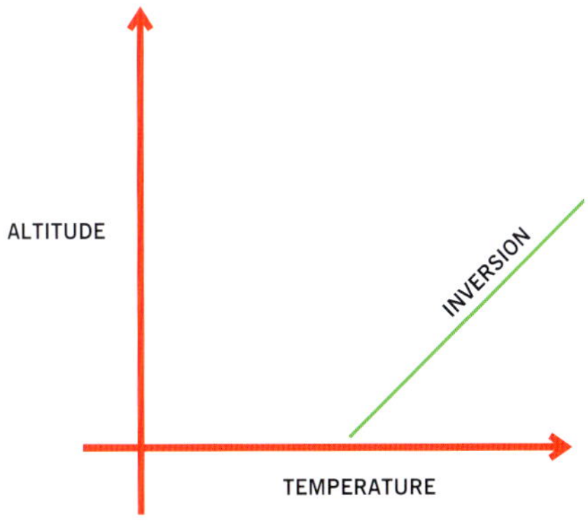

°C	°F[1]	°C	°F[1]
-40	-40.0	+40	+104.0
-39	-38.0	+39	+102.0
-38	-36.5	+38	+100.5
-37	-34.5	+37	+98.0
-36	-33.0	+36	+96.5
-35	-31.0	+35	+95.0
-34	-29.0	+34	+93.0
-33	-27.5	+33	+91.5
-32	-25.5	+32	+89.5
-31	-24.0	+31	+88.0
-30	-22.0	+30	+86.0
-29	-20.0	+29	+84.0
-28	-18.5	+28	+82.0
-27	-16.5	+27	+80.5
-26	-15.0	+26	+79.0
-25	-13.0	+25	+77.0
-24	-11.0	+24	+75.0
-23	-9.5	+23	+73.5
-22	-7.5	+22	+71.5
-21	-6.0	+21	+70.0
-20	-4.0	+20	+68.0
-19	-2.0	+19	+66.0
-18	-0.5	+18	+64.5
-17	+1.5	+17	+62.5
-16	+3.0	+16	+60.0
-15	+5.0	+15	+59.0
-14	+7.0	+14	+57.0
-13	+8.5	+13	+55.5
-12	+10.5	+12	+53.5
-11	+12.0	+11	+52.0
-10	+14.0	+10	+50.0
-9	+16.0	+9	+48.0
-8	+17.5	+8	+46.5
-7	+19.5	+7	+44.5
-6	+21.0	+6	+43.0
-5	+23.0	+5	+41.0
-4	+25.0	+4	+39.5
-3	+26.0	+3	+37.5
-2	+28.5	+2	+35.5
-1	+30.5	+1	+33.5
0	+32.0	0	+32.0

Figure M2.6

A temperature conversion scale.

[1]Fahrenheit temperatures are rounded to the nearest 0.5 degree.

In aviation meteorology, temperature is usually referred to in terms of degrees 'Celsius', which is the same as degrees 'Centigrade'. On this scale, water freezes at 0°C and boils at 100°C. Temperature may also be referred in degrees 'Fahrenheit', a reference which remains common in some European countries and the USA. To convert from Celsius to Fahrenheit there are a couple of methods:

To convert °C to °F divide by 5, multiply by 9, and add 32;

eg 15°C ÷ 5 (= 3); x 9 (= 27) + 32 = 59°F

To convert °F to °C subtract 32; divide by 9, multiply by 5;

eg 59°F – 32 (= 27); ÷ 9 (= 3); X 5 = 15°C

For those of us who do not excel in mental arithmetic, there is a quicker, if slightly less accurate, rule of thumb. Start from the basis that 10°C = 50°F. Thereafter allow 9°F for each 5°C:

For example, 20°C = 68°F.

This rule can also be used in reverse:

40°F = 4°C.

To avoid too much mental arithmetic, graphs and conversion tables (such as that on a mechanical flight computer) are widely available. For most non-mathematicians, these give a better chance of arriving at an accurate conversion between different units of temperature.

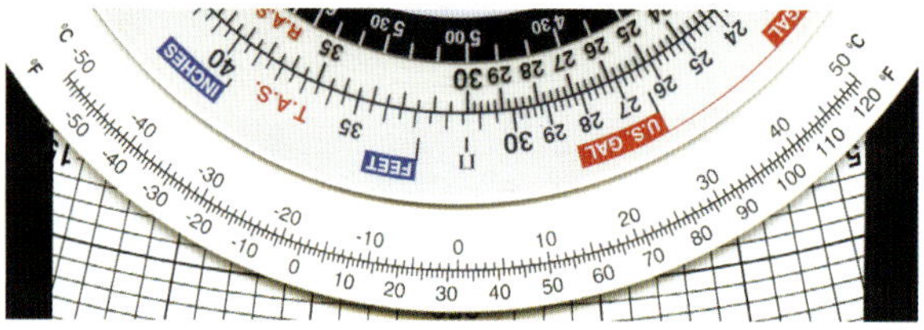

Figure M2.7

The temperature conversion scale on an ARC mechanical flight computer

Pressure variation in the atmosphere

Atmospheric **pressure** is a factor that pilots will consider before and during every flight (not least because the aircraft's altimeter is set by reference to the current atmospheric pressure; at least during take-off and landing).

As surface-dwelling creatures (most of the time!), we humans live at the bottom of the ocean of air that surrounds the planet, and due to gravity we are subject to the weight of the entire column of the atmosphere pressing down on us. This creates an average sea level pressure of around 14 pounds per square inch (psi), or about 1 'bar' in metric terms. This is equal to a weight of just over ten tonnes per square metre. Of course, because we have evolved over millions of years to survive at this pressure, we don't notice it. In fact, we rely on this intense pressure to be able to breathe.

Nevertheless, the air pressure at any particular point will have a profound effect on the properties and behaviour of the atmosphere, a feature complicated by the fact that air pressure across the planet's surface is constantly changing. A large area of low pressure may be referred to as a **depression**. A large area of high pressure may be referred to as an **anticyclone**. Such areas of low and high pressure will typically cover areas of hundreds of thousands of square miles (or an even greater number of square kilometres, if you prefer metric units).

Pressure also varies with altitude. As altitude increases, there is less weight of the atmosphere pressing down and so air pressure reduces.

The pressure of the atmosphere can be expressed in several different units. In aviation the most common unit of atmospheric pressure is the **hectopascal** (hPa) – formerly known as the 'millibar' (mb). The other well-known unit of pressure measurement is 'inches of mercury', as employed on the dials of antique barometers but also used for aviation purposes in some countries, notably North America. Many flight guides carry a conversion table between hectopascals and inches of mercury. But there again, a few Eastern European countries express atmospheric pressure in millimetres of mercury...

The reduction of pressure with increasing altitude does not take place at a constant rate. At the surface, pressure reduces at about one hPa per 27ft. By 20,000ft the rate of pressure reduction is about one hPa per 50ft, and by 30,000ft it is about one hPa per 75ft.

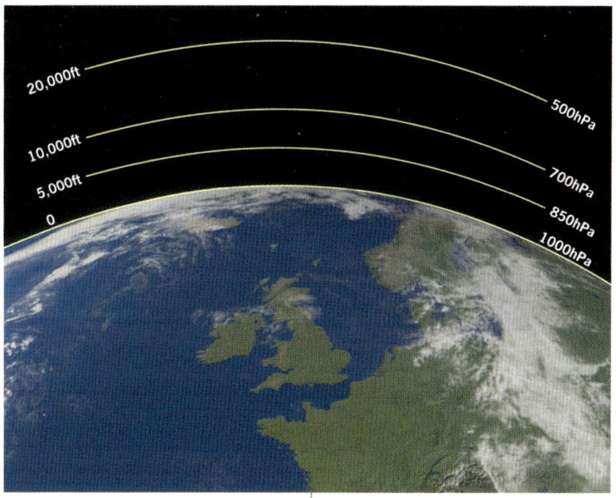

Figure M2.8
The reduction of pressure with increasing altitude does not take place at a constant rate.

Density

Air density is a measure of the mass of the air in a given volume of air. Although pilots rarely need to know the actual density of a parcel of air, the common unit of measurement of air density is kilograms per cubic metre.

Air density is mostly dictated by temperature and pressure.

Air density is inversely proportional to temperature. As air temperature increases, air density reduces (and as air temperature reduces, air density increases). So, the hotter the air, the less dense it is.

Air density is proportional to pressure. As air pressure increases (is compressed), air density also increases. And as air pressure reduces (air expands), air density reduces.

From measuring the atmosphere, meteorologists know that as altitude increases, air density reduces. But this leaves a puzzle. Reduced air pressure (such as happens with increasing altitude) will reduce air density. But a reduction in air temperature (which also happens with increasing altitude) implies an increase in density. So what is happening? In fact, the change in pressure has a greater effect on density than the change in temperature; so as altitude increases, air density does indeed reduce.

Localised differences in air density are one factor that can lead to vertical movement of air in the atmosphere. Most people will have noticed that in a fizzy drink, bubbles of less dense gas rise up through the more dense liquid. In the same way, bubbles of less dense air will tend to rise through an area of more dense air.

Humidity

It has already been stated that two of the principal properties of the atmosphere are temperature and pressure. The third major property is water content.

The amount of water vapour (which you will remember, of course, is an invisible gas) in the atmosphere is referred to as absolute **humidity**. The greater the proportion of water vapour in the atmosphere, the higher the humidity. This is important as the formation of almost all weather phenomena (clouds, fog, rain, snow, ice etc.) is dependant on the amount of water vapour in the atmosphere.

A given quantity of air can contain a certain amount of water as invisible water vapour. The proportion of water vapour which a parcel of air can hold varies with a number of factors, principally temperature. Warm air can hold more water vapour than cold air at the same pressure. By way of illustration, at 0°C a kilogram of air can hold up to 4g of water vapour. At +15°C, a kilogram of air can hold up to 10g of water vapour, and at +30°C it can hold up to 27g of water vapour.

The question is, what happens when there is too much water vapour for the air to hold? Imagine a 1kg parcel of air with a temperature of +25°C. It is holding 17gm of water vapour. If this parcel of air cools, the amount of water vapour it can hold decreases. Eventually (at about +21°C in standard atmospheric conditions) the parcel reaches the limit of its ability to retain its water content as water vapour. At this temperature, it is holding all the water vapour it possibly can, and is said to be **saturated**. If the parcel of air is cooled any further, the water vapour will begin to condense into liquid and become visible as water droplets suspended in the air – otherwise known as **cloud**.

Figure M2.9

As a parcel of air cools, the amount of water vapour it can hold decreases. If the parcel of air is cooled below its 'dew-point' temperature, the water vapour will condense into water droplets – a cloud.

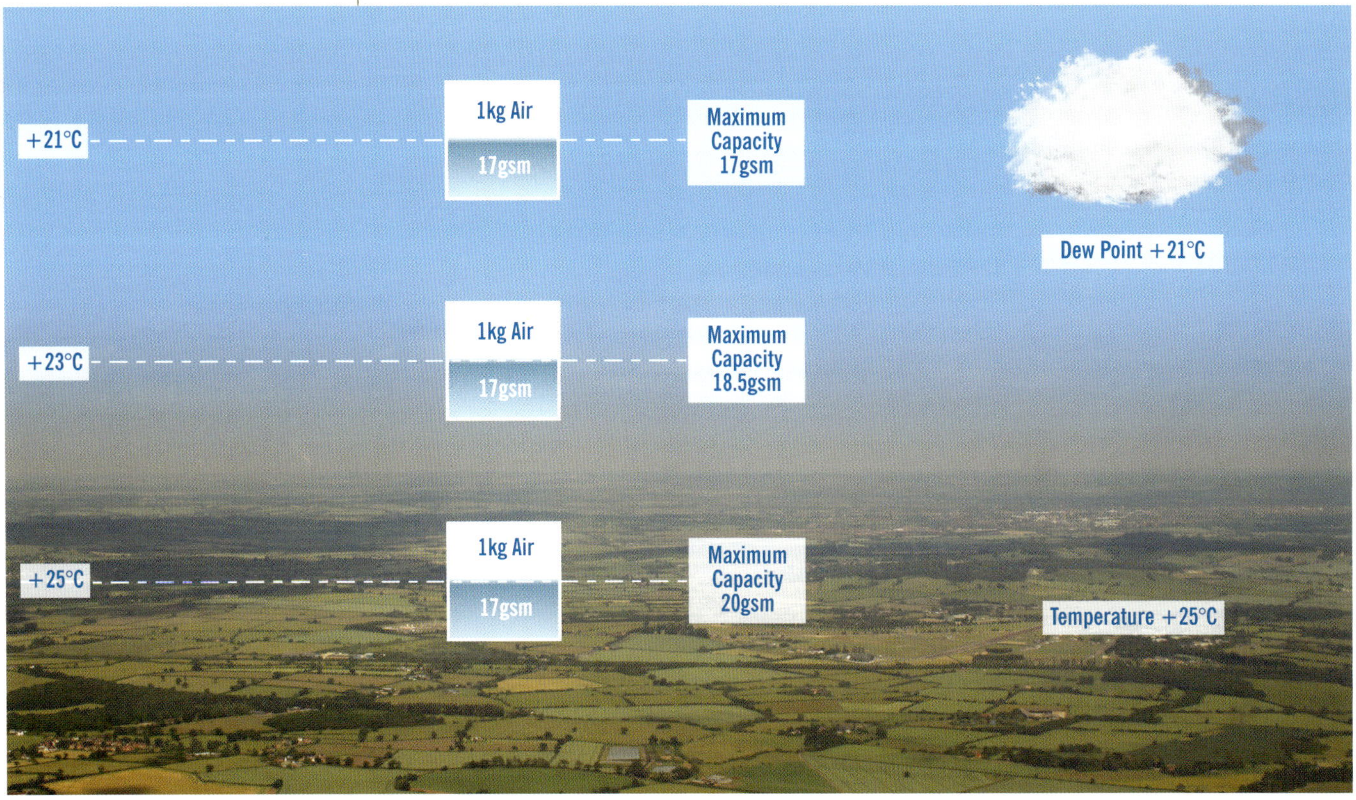

The temperature at which the air will become saturated (ie it is holding the maximum amount of water vapour it can without that vapour condensing into water droplets) is known as the **dew point** temperature. Dew point is measured in meteorological observations and reported in actual weather reports.

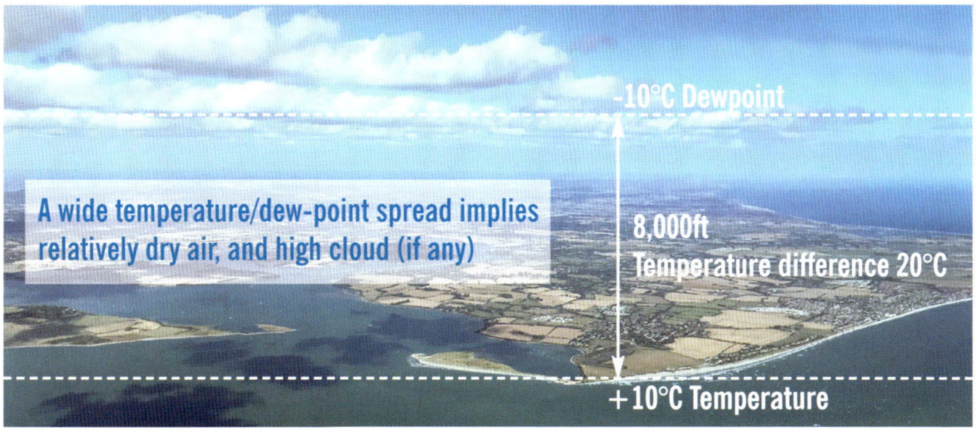

Figure M2.10
A wide spread between surface air temperature and dew-point implies a high cloud base.

Figure M2.11
A narrow spread between surface air temperature and dew-point implies a low cloud base.

The relationship between dew point temperature and actual temperature gives a good indication of the percentage of water vapour in the air. Where there is a large gap between temperature and dew point, the air is quite dry. In this case the air would have to be significantly cooled, or have a lot of moisture added to it, for condensation to occur and cloud to form. A 'wide' temperature/dew point spread implies a high cloud base, or no cloud at all.

If there is a small gap between temperature and dew point, the air is much closer to saturation and just a small reduction in temperature, or a little additional moisture, could cause cloud to form. A 'narrow' temperature/dew point spread implies the possibility of low cloud or even fog (which is just cloud on the surface).

The percentage of water vapour a parcel of air is holding, compared to the amount it could hold before becoming saturated, is known as the **relative humidity**. In other words, if a parcel of air contains half the water vapour it could hold before becoming saturated, the relative humidity is 50%. If air is saturated the relative humidity is 100%, because the air is holding all the water vapour it can before condensation will occur. We have seen that warm air can hold more water vapour than cold air. So warm air with 90% relative humidity contains more water vapour than cold air with 90% relative humidity.

The weather information given to pilots does not include the relative humidity, but relative humidity is of great interest to meteorologists in weather forecasting. In computer weather forecasting models, the three most important atmospheric measurements are those of pressure, temperature and humidity.

A final point for now regarding humidity. Water vapour is less dense than dry air. Therefore, **moist air** (high humidity) **is less dense than dry air** (low humidity).

The International Standard Atmosphere (ISA)

In many ways, the atmosphere behaves much like a fluid in motion and basic parameters of temperature, density and pressure are constantly changing at any one location, let alone at different locations. This could make life very difficult for those who need a constant calibration to work to – for example instrument manufacturers, aircraft designers and so on. To overcome this problem, an 'average' set of atmospheric conditions, known as the **International Standard Atmosphere** (**ISA**) has been established.

Altitude	ISA Temperature
10,000ft	-5°C
8000ft	-1°C
6000ft	+3°C
4000ft	+7°C
2000ft	+11°C
Sea Level	+15°C

Figure M2.12
Temperatures in the International Standard Atmosphere.

The basic parameters of the ISA are quite simple. At sea level, the following conditions apply:

- Temperature +15°C
- Pressure 1013·25hPa
- Temperature Lapse Rate 1·98°C/1000ft. The temperature decreases at this rate until 36,090ft above sea level, above which it remains at a constant -56·5°C

Tables of the International Standard Atmosphere allow temperature and pressure values to be obtained for different levels in the atmosphere. Nevertheless, using a rounded lapse rate of 2°C/1000ft, working out the ISA temperature at altitude is not a difficult calculation even for a non-mathematician.

Of course, actual atmospheric conditions are invariably different from those of the ISA. If the actual air temperature (sometimes referred to as the **Outside Air Temperature** or **OAT**) at 3000ft is +10°C, whereas the ISA temperature at 3000ft is +9°C, the actual temperature can be described as *"ISA +1"*. Performance tables for an aircraft may be marked 'ISA'; 'ISA +10°C'; 'ISA -10°C' etc. to allow aircraft performance to be calculated at differing air temperatures. OAT can also be a limitation in a flight manual eg *"Flight in conditions beyond ISA +30°C not permitted"*.

The difference between ISA conditions and the ambient (actual) conditions can have a marked effect on aircraft performance and the accuracy of aircraft instruments. By way of illustration, imagine an average light aircraft which in ISA sea-level conditions has a take-off distance of 650 metres. If the same aircraft is taken to Samedan airfield in Switzerland, (elevation 5600ft), when the temperature is +25°C, its take-off distance required is likely to be around 1100m – almost double!

Progress Check

5. On average, how does air temperature change as altitude increases?
6. What is a temperature inversion?
7. How does air pressure change as altitude increases?
8. What name can be given to a large area of low pressure?
9. In European aviation meteorology, what unit is used to measure pressure?
10. How does air density change with increasing altitude?
11. If pressure remains constant, but temperature is increased, what is the effect on density?
12. Can cold air hold more, or less, water vapour than warm air?
13. What name is given to the temperature at which air will become saturated?
14. What is the sea level pressure in the International Standard Atmosphere (ISA)?
15. What is the temperature at 5000ft Above Mean Sea Level (AMSL) in the International Standard Atmosphere (ISA)?

These questions are intended to test knowledge and reinforce some of the key learning points from this section. In answering these questions, a 'pass rate' of around 80% should be the target.

Model answers are found at page M172.

M3 Altimetry

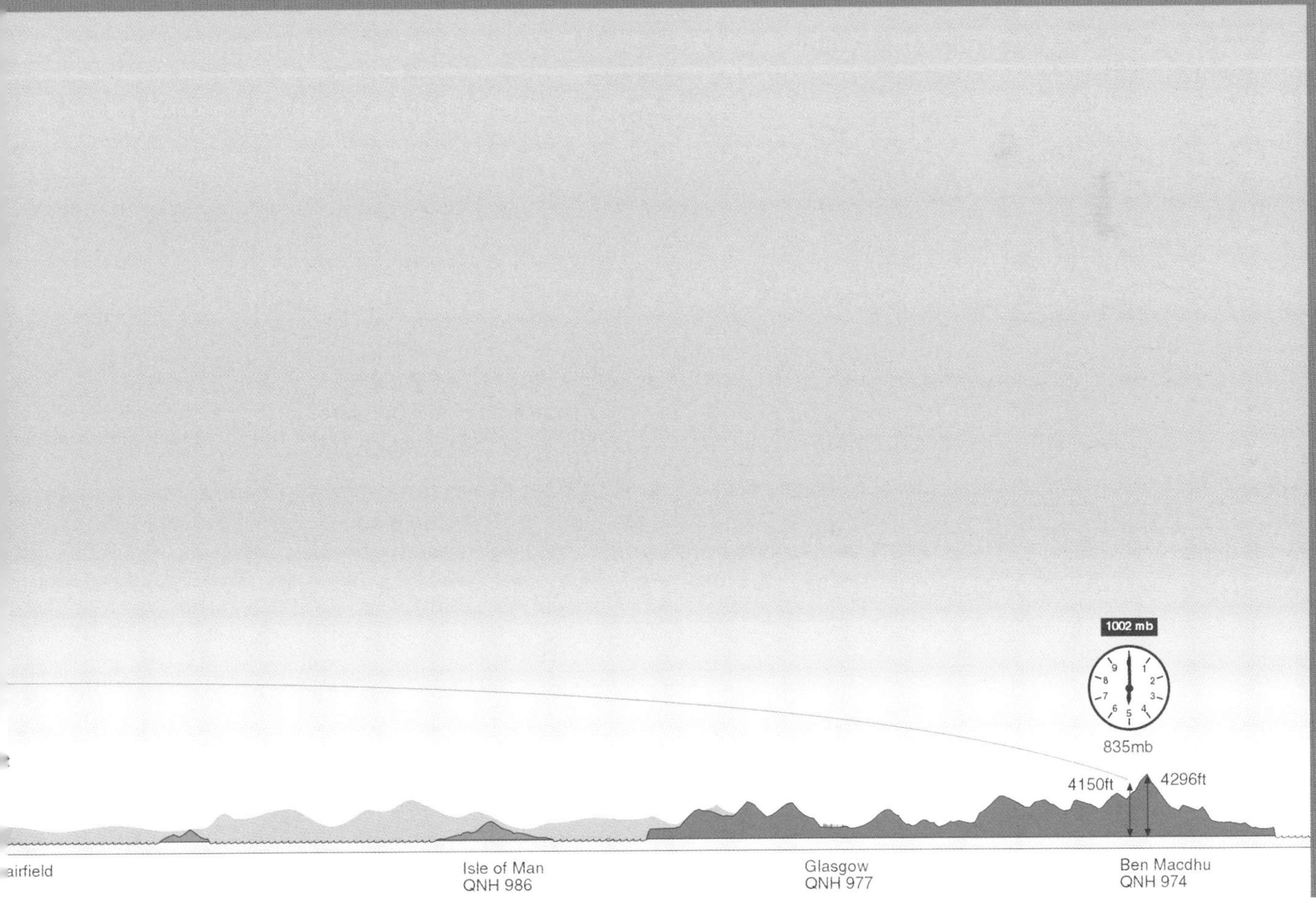

M3 Altimetry

Altimeter and Pressure Settings

Figure M3.0
A barometer measures atmospheric pressure.

Meteorologists measure **atmospheric pressure** using a sensitive barometer, which in Europe usually indicates the observed pressure in hectopascals (although older barometers may be calibrated in inches of mercury).

Every aircraft also carries a sensitive barometer – it is called an **altimeter**. The altimeter is just a form of barometer which consists of a set of metal 'bellows' that have been partially evacuated and sealed. As air pressure rises and falls the bellows expand and contract. This movement is measured, calibrated, and displayed on the altimeter face in terms of vertical distance.

This raises an important point, which is well-worth emphasising. A normal altimeter in a General Aviation (GA) aircraft does **not** display true vertical distance above the surface. In a GA aircraft an altimeter is merely a barometer that displays the atmospheric pressure in the form of vertical distance above a datum selected by the pilot.

Imagine for a moment being at the seaside. You have an altimeter with you and set the altimeter sub-scale to the observed atmospheric pressure, which we will say is 1000hPa. This is a pressure setting known as **QNH** – the atmospheric pressure at sea level. When the correct QNH is set on the altimeter the altimeter will display the vertical distance above sea level. If the correct QNH is set, and if the altimeter is at sea level, it should read zero.

With the QNH set on the altimeter sub-scale, the altimeter will indicate **altitude** above the average level of the sea, and this level may be referred to as '**Above Mean Sea Level**' (**AMSL**).

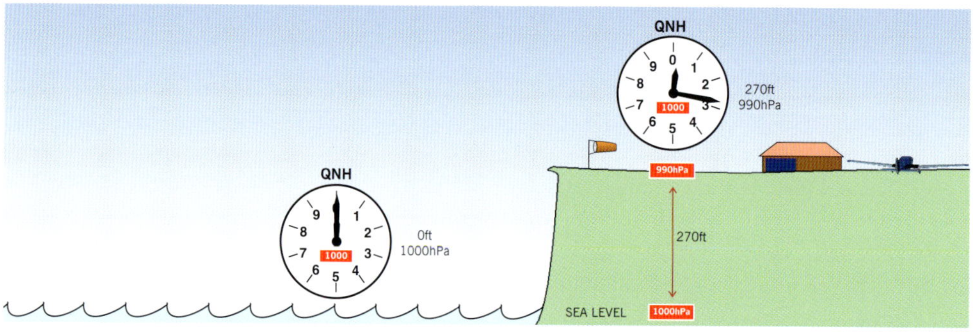

Figure M3.1
If the sea level air pressure is 1000hPa, an altimeter set to 1000hpa (the QNH) will read zero at sea level. Raise that altimeter, still with the 1000hPa pressure setting, to a point where the air pressure is 990hPa, and it will display a vertical distance of 270ft.

Imagine now taking the altimeter a short distance to an airfield that has an elevation (its vertical distance above sea level) of 270ft. In 'standard atmospheric conditions', pressure reduces by around 1hPa for each 27ft increase in altitude in the lower atmosphere. It follows that at the airfield the altimeter will now be measuring a pressure of about 990hPa (10hPa less than the pressure at sea level) and it will display this measured pressure as an altitude of around 270ft. In other words, it is displaying the elevation (or altitude) of the airfield. The airfield QNH is 1000hPa.

QNH is one of three altimeter settings.

If you want the altimeter to display vertical distance against some specific point on the surface other than sea level, a pressure setting known as **QFE** is required. When QFE is set on the altimeter, the **height** above a specific point (usually an airfield) is displayed on the altimeter. In the example given above, if the QNH is 1000hPa and the airfield is 270ft above sea level, the QFE pressure setting will be around 990hPa. If you change the altimeter sub-scale setting to 990hPa, the altimeter should read zero on the airfield. This is because the altimeter is now set to the airfield QFE, and it is reading height above a specific datum – which in this case is the airfield itself.

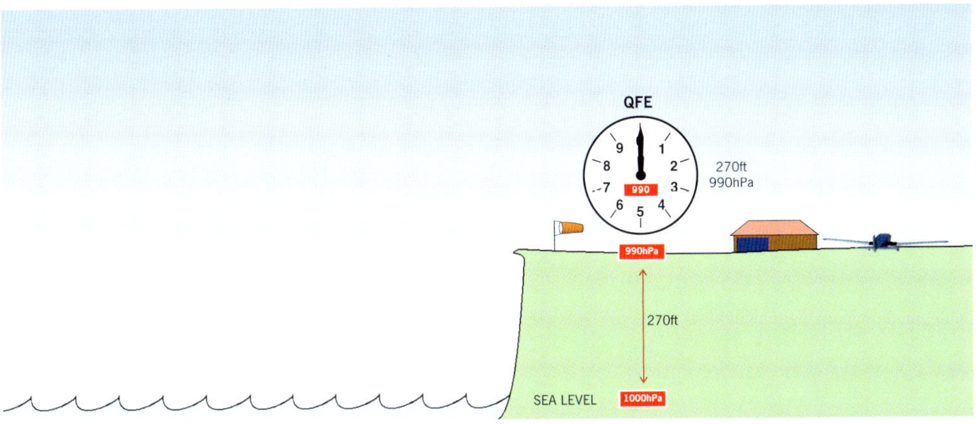

Figure M3.2
If the surface pressure at an airfield is 990hPa, an altimeter set to 990hpa (the QFE) will read zero on the airfield.

The whole point of this theoretical exercise is to demonstrate the fallibility of the altimeter. Clearly, taking the aircraft's instrumentation. to the seaside before going flying is not a practical proposition; instead pilots rely on the reported QNH at an airfield to tell them the sea-level pressure. But if pressure is reported wrongly, or if a pilot sets the wrong pressure on the altimeter sub-scale, the altimeter is no longer giving an accurate reading. This is why it is **vital** to use the correct setting on the altimeter sub-scale, because **the altimeter can only indicate the vertical distance above whatever datum the pilot has selected.**

It is clearly of crucial importance to be absolutely clear which datum is being used on the altimeter and to set the correct pressure setting on the altimeter sub-scale. Because by flying at a constant indicated altitude, a pilot is in fact following a level of constant pressure. It is also vital that the altimeter setting is periodically updated. To illustrate the importance of this point, let's go on an imaginary flight.

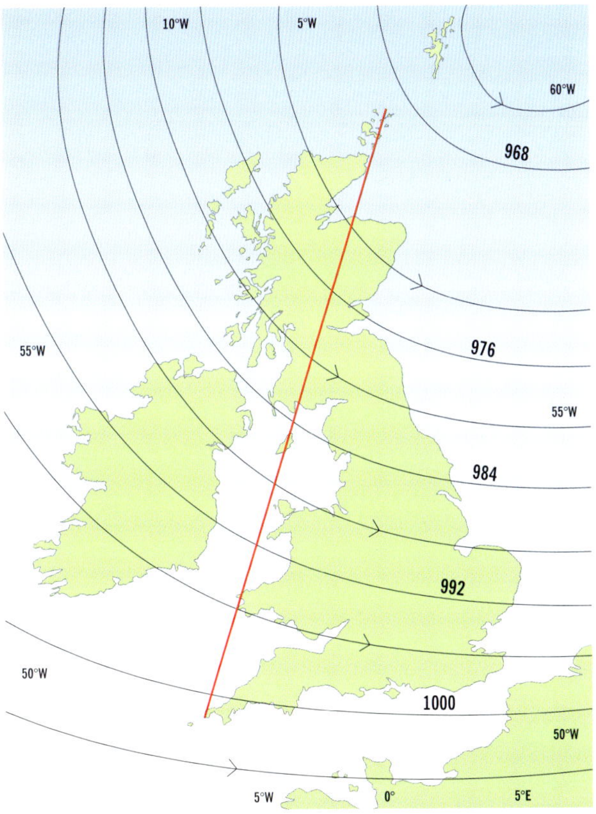

Figure M3.3
The pressure distribution across the UK on the day of an imaginary flight from Lands End to Kirkwall.

Imagine a situation where there is a steep 'pressure-gradient' across the United Kingdom (in other words, there is a significant change in atmospheric pressure over a relatively short distance). The sea-level pressure (QNH) around the Land's End area is 1002hPa, and a pilot is planning to fly from there to Kirkwall in the Orkneys where the QNH is 969hPa. Being a busy person, this pilot draws a straight line from Land's End to Kirkwall; notes that the highest bit of ground en route is at about 4200ft and decides that 5000ft will be a safe altitude (please note: this careless attitude to flight planning is not recommended). After take-off the intrepid aviator climbs to 5000ft, sets the autopilot to 'Altitude Hold' and heads north.

For the purposes of this example, and to keep the mental arithmetic to a minimum, let's assume that 1hPa of pressure change can be taken to equal a 30ft change in altitude up to about 10,000ft altitude. So at an indicated 5000ft, the aircraft will on this day be at a pressure level of about 835hPa; that is 1002 – (5000ft ÷ 30ft per hPa). As long as the aircraft stays at the pressure level of 835hPa, the altimeter will read 5000ft. The problem arises if the altimeter setting is not updated. Flying towards an area of low pressure, the 835hPa pressure level is lowering, coming closer to the surface.

Let's assume that our intrepid pilot is not updating the altimeter setting. Passing abeam Snowdonia, where the QNH is about 990hPa, the 835hPa pressure level is about 4640ft above sea level (the difference between the actual QNH and that set on the altimeter is 12hPa x 30ft = 360ft); even though the altimeter still reads 5000ft. But our pilot is still over the sea, and still more than 1000ft higher than the highest point in Snowdonia. By the time our hapless pilot is passing over the Isle of Man, where the sea level pressure – QNH – is about 986hPa, the aircraft is down to 4520ft Above Mean Sea Level (AMSL) – but with the altimeter still reading 5000ft. The highest point on the Isle of Man is about 2200ft AMSL, so our pilot is still in luck, but the aircraft is getting lower and lower all the time. Remember, the altimeter is still reading 5000ft.

If our aviator was thinking clearly, alarm bells might be ringing by now. The flight will make it over the Scottish border, but passing between Glasgow and Edinburgh (where the QNH is about 977hPa) the aircraft will now be down to about 4250ft. However, with the Land's End QNH of 1002 still set on the altimeter, the pilot still sees 5000ft indicated on the altimeter.

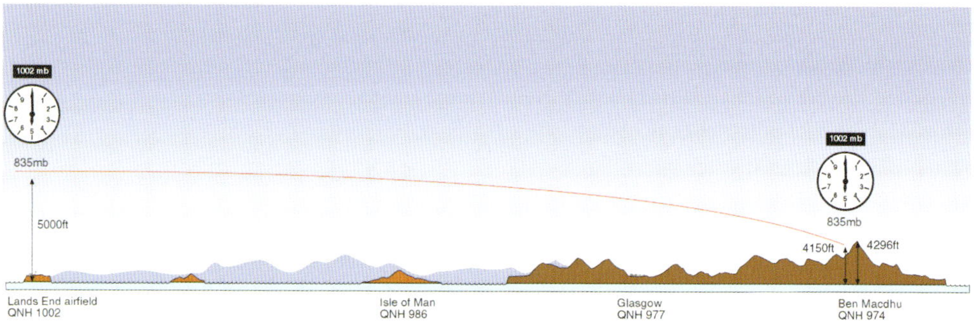

Figure M3.4
Flying towards an area of low pressure, if the altimeter setting is not updated, the aircraft will descend (following a constant pressure level) even though the altimeter shows a constant level.

Of course, the altimeter is working properly. It is indicating that the aircraft is 5000ft above the 1002hPa pressure level, which is true. However; by now the theoretical 1002hPa pressure level is some way below ground. This flight is likely to terminate rather violently in the Cairngorm mountains, probably on the southern slope of Ben Macdui just below the summit (which is at 4296ft AMSL). This very beautiful – if somewhat desolate – spot is a full 117 nautical miles short of the destination – Kirkwall. The accident investigator will note that the crash site is about 4150ft above sea level. The investigator will also note that the altimeter was reading 5000ft at impact...

Do not think that this is an extreme example, it is not. The pressure gradient used as an example is taken from a real weather chart and is by no means exceptional in Europe in the winter.

The point is that on a long-distance flight, even a less steep pressure-gradient will lead to significant inaccuracies in altimeter reading if the altimeter pressure setting is not updated regularly. The result may not be as dramatic as that imagined above, but it could still be the cause of a close encounter with another aircraft, or an infringement of controlled airspace.

In the example above, flying from a high-pressure area to a low-pressure area, the altimeter will **over-read** and indicate that the aircraft is higher than it actually is, unless the sub-scale setting is regularly updated.

Hence the moto: **high to low, down you go**.

So far we have focussed on the importance of setting the correct pressure setting on the altimeter sub-scale – but what if you don't know the correct pressure setting? There are so many ways to obtain current (and forecast) weather information, that it's difficult to see such a situation occurring. Nevertheless, if you have understood the principle of operation of the altimeter, you will realise that if you are on the ground at an airfield and set the altimeter to read the airfield elevation (its altitude above sea level), the pressure setting in the sub-scale window is the airfield QNH. In the same situation, if you set the altimeter to read zero, the pressure setting in the sub-scale window is the airfield QFE.

In any event, when you are given a pressure setting before flight, setting it on the altimeter and checking the altimeter indication in relation to the airfield elevation (to check QNH) or against zero (to check QFE) is a good way to cross-check the pressure setting and the altimeter accuracy.

To summarise these two altimeter pressure settings:

Pressure Setting	Description	Terminology
QNH	Altimeter reads vertical distance Above Mean Sea Level (AMSL)	Altitude
QFE	Altimeter reads vertical distance above a specific point on the surface	Height

Figure M3.5
A summary of altimeter pressure settings.

Incidentally, the origins of most 'Q' codes are lost in the mists of time, but some pilots claim that QFE stands for **Q F**ield **E**levation, which is one way of remembering its meaning.

Altimeter Temperature and Pressure Effects

Like many aircraft instruments, the aircraft altimeter is calibrated for International Standard Atmosphere (ISA) conditions, which include a sea level temperature of +15°C and a temperature lapse rate of 1.98°C per 1000ft. It follows that if the actual temperature is significantly warmer or colder than the ISA temperature, some form of error will be introduced into the altimeter reading, even if the correct pressure is set. Not surprisingly, errors in altimeter readings caused by non-standard temperatures are known as **temperature error**.

Warm air is less dense than cold air. So given two columns in the atmosphere containing a fixed amount of air, the warm column will be taller than the cold column even if they both have the same surface pressure. It follows that any given pressure level will be higher in the warm air column than in the cold air. In air which is warmer than standard, the altimeter will **under-read** and the aircraft is **higher** than indicated. In air which is colder than standard, the altimeter will **over-read** and the aircraft is **lower** than indicated. So, if flying from an area of warm air into an area of cold air, the change in temperature will cause the aircraft to be lower than indicated on the altimeter:

**High to Low,
Down you go.**

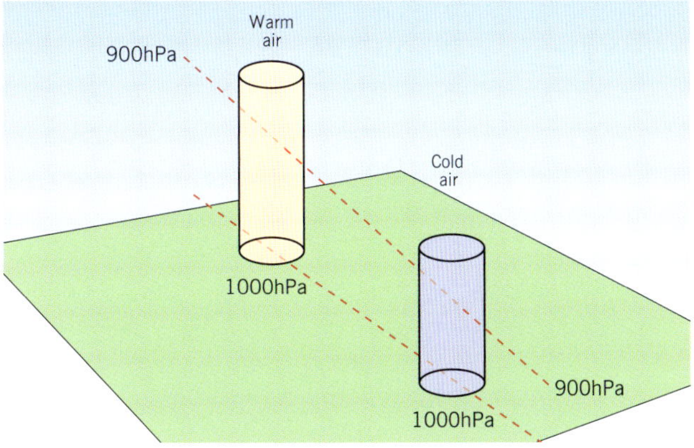

Figure M3.6
In warm air, any given pressure level is higher than it is in cold air.

Figure M6.3
Advection fog over a large lake.

Advection fog over the sea is most common during the spring and summer; over land it tends to be quickly dispersed by daytime heating and should not persist too far inland from the coast, although poor visibility and low cloud may be a problem along the coast itself. That said, low cloud and advection fog can penetrate quite far inland overnight, particularly along river valleys. Thus airfields in valleys which are open to the sea can be particularly vulnerable to low stratus and fog moving in from the sea, which reaches them earlier than other airfields, and stays longer.

Advection fog over the sea does not vary between day and night, and it will tend to persist until a different airflow arrives. By its nature, this type of fog is relatively simple to predict and should not catch out a pilot who has been checking weather forecasts and reports.

Radiation fog, on the other hand, can be much more localised and far less predictable than advection fog. Radiation fog forms over the land and is caused by the surface cooling at night.

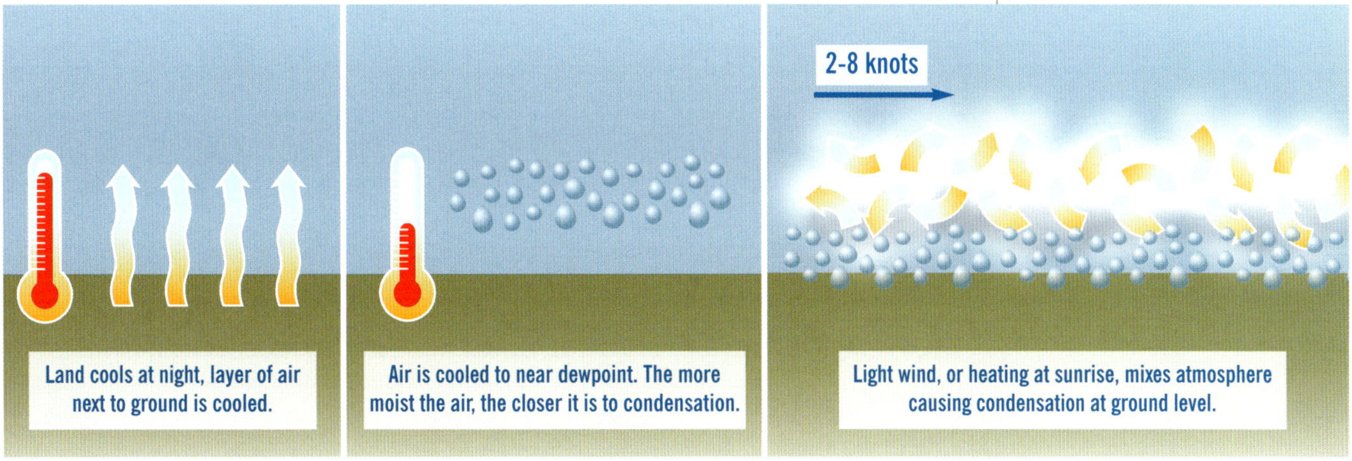

Figure M6.4
The three criteria for radiation fog:

- cooling of the land surface at night;
- air cooled to its dew-point;
- light winds to 'mix-up' the atmosphere at the surface.

On a clear night the ground loses heat to space, a process known as radiation (hence the term 'radiation fog'). A night sky which is cloud-free or nearly so is important in this respect because cloud tends to act as an insulator, retaining warmth when the sky is covered by overcast cloud at night. Without cloud cover, night-time cooling of the land surface is far more pronounced, and the layer of air next to the surface will also be cooled by contact with it.

Once this cooling process is under way, the water vapour content of the air is all-important. The greater the relative moisture content of the air (the higher its relative humidity) the more easily the air will reach saturation. When conditions are favourable for the formation of radiation fog, meteorologists and experienced pilots watch the temperature and dew-point closely. The closer they become (the narrower the temperature/dew-point spread) the greater the likelihood of radiation fog. As a general rule, once the temperature/dew-point spread is less than 2°C and becoming more narrow, fog or very low cloud may be imminent.

The final factor in the formation of radiation fog is the wind. Light winds – anywhere between about 2 and 8 knots – are ideal for the creation of radiation fog. If the wind speed is less than 2 knots, a heavy dew will probably form instead of fog (although occasionally fog may form even in calm conditions). Much more than 10 knots wind speed and the increased mixing in the atmosphere will tend to cause low cloud to form rather than ground-based fog.

To summarise, the three principal factors for the formation of radiation fog are:

- Little or no cloud at night and cooling of the land surface by radiation;
- High relative humidity;
- Light winds, ideally 2-8 knots.

Radiation fog may start to appear around sunset, but statistically it is most likely to occur in the early hours and around dawn. The most common time of year for the occurrence of radiation fog in Europe is the late autumn, when the diurnal (daily) variation of temperature is greatest, although the spring sees a lesser peak in fog occurrence. An anticyclone or ridge bringing clear skies and light winds, and southerly or southwesterly winds tend to be associated with air with the highest relative humidity. But none of these can be taken as absolutes.

Figure M6.5

Early morning radiation fog.

Figure M6.6

An airport shrouded in early-morning fog.

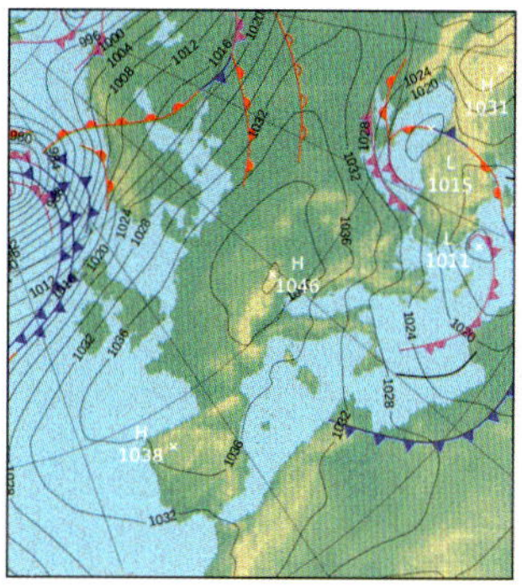

Figure M6.7

An established area of high pressure over Europe in winter is part of the ideal set-up for radiation fog.

It is the fine balance of atmospheric conditions capable of causing radiation fog that often make it so unpredictable. During a particularly calm night, it is possible that little or no fog may form even though all other conditions are favourable. However, at sunrise the heating of the surface – leading to convection currents which stir up the atmosphere – can cause enough atmospheric mixing for thick fog to form rapidly just after dawn.

Figure M6.8

Radiation fog persisting in deep river valleys still in shadow.

Terrain can also have a marked effect on the formation of radiation fog. Radiation fog tends to form earliest, and last longest, in valleys. In valleys, cold air sinks ('katabatically') to the valley floor at night, where there may be a river or lake to increase the relative humidity. Once the sun rises in the morning, the valley may stay in shadow for longer than elsewhere, denying the sun's heat the chance to disperse the fog. However, an airfield on a hill is not guaranteed to stay fog-free. Fog may form below the airfield overnight, leaving it in the clear. However, after dawn, the fog may be warmed and lift to cover the airfield, just as everywhere else is becoming clear of fog!

EGHH BOURNEMOUTH AIRPORT

METAR EGHH 020850Z 36004KT 0150 FG ... 07/07

Figure M6.9

An extract from a METAR report from Bournemouth airport at 0850 UTC in the morning. The wind is light (360/04); the visibility is 150 metres in fog and the temperature and the dew-point are both +7°C.

Dispersal of radiation fog (or 'burning-off', as pilots and meteorologists often describe it) usually takes place as the sun increases the ground temperature, thus heating the air and causing the fog to evaporate (water droplets becoming water vapour). It follows that radiation fog in the summer (which is quite rare) tends to be quickly dissipated by the strong sun. In winter the sun may not be strong enough to disperse the fog completely, and it may then linger in places throughout the daylight hours and then become widespread again at night. This cycle can persist for several days. Radiation fog can be cleared by strengthening winds, which initially lifts the fog into a layer of low stratus before breaking it up.

Figure M6.10
In the early morning this high ground is above surrounding radiation fog, but the fog may soon lift into low cloud and cover the high ground.

Figure M6.11
The sun has just become visible through the fog (top left) and has begun to warm the land. This summer morning fog soon lifted and dispersed .

The clearance of fog is often a difficult topic between weather forecasters and pilots, because it is very difficult for a forecaster to get the timing of fog clearance correct. The forecasting of fog clearance is based on a precise knowledge of the conditions throughout the atmosphere at each place where fog has formed. Unfortunately, this information can never be fully available to the forecaster, who has to make an educated guess of the time at which the fog will clear. It may be a statistical freak, but most pilots would agree that the forecast times of fog clearance are more often than not over-optimistic.

Figure M6.12
Seen from above, radiation fog lifting into broken low cloud.

Regrettably it is not uncommon for the predicted clearance time of fog clearance to be put back repeatedly as the fog fails to clear when it was supposed to. A weather forecast with a predicted fog clearance time is like a cigarette packet – it should carry a prominent health warning! If you are relying on a forecast fog clearance in order to reach an airfield, remember that the clearance may not happen at all, let alone at the forecast time. Fog clearance also varies in different areas. If flying to an airfield without meteorological reporting facilities, do not assume that it will be fog-free just because a nearby airport is clear. A quirk of the local terrain (maybe the destination airfield is in a valley or next to a lake) could easily leave the destination fog-bound. Caution is required in these conditions – wise pilots will carry plenty of reserve fuel and have fog-free alternates planned.

Apart from the two principal types of fog, there are a few others – rarer but worth mentioning.

Steaming fog forms above water when the air is much cooler than the water. It is a phenomenon sometimes seen when a water-covered road steams in strong sunshine – or if you put a bowl of warm water into a refrigerator. Although relatively rare in most of Europe, it can form quite easily in the Norwegian fjords. It is common in polar and arctic regions and is also known as 'arctic sea smoke' or 'frost smoke'.

Frontal fog occasionally forms just ahead of a warm or occluded front (described later), either as the cloud comes down to ground level or if the air becomes saturated by continuous rain.

You may also come across the term 'hill fog'. This merely describes cloud covering high ground – because fog is nothing more than cloud at ground level. 'Hill fog' can also be described as 'cloud covering the hills'.

Haze and Smoke

When visibility is reduced by smoke or particles of solid matter, a number of different descriptions can be used to describe the weather phenomenon. Collectively, they are known to weather forecasters (who must have a bigger dictionary than most pilots) as 'obscuration'.

Figure M6.13
A strong inversion can act as a 'lid' on the lower atmosphere, trapping smoke, dust and haze in the inversion layer.

5230N 05E			
24	020	10	-34
18	020	10	-23
10	240	10	-04
05	240	15	+06
02	250	20	+10
01	240	15	+00

Figure M6.14
In this 'spot wind' forecast, a temperature inversion is evident as the temperature increases markedly between 1000ft and 2000ft – increasing from 0°C at 1000ft to +10°C at 2000ft.

Haze (HZ) The term 'haze' is often used as a catch-all description of smoke and dust in the atmosphere and is frequently associated with anticyclonic (high pressure) conditions. The subsiding (sinking) air of an anticyclone tends to encourage one or more temperature inversions to form. An inversion acts as a 'lid' in the atmosphere, forming a haze layer which usually reaches from the surface up to the inversion. The longer the anticyclone persists, the worse the visibility becomes as smoke and other pollutants are pumped into the atmosphere and trapped under the inversion. The top of the haze layer (usually at the inversion) is often sharply defined and flying conditions above it are usually clear and smooth. However, flight in and above haze presents some potential problems for the unwary.

Figure M6.15
A marked haze layer, with markedly poorer visibility within the haze layer.

Figure M6.16
A distinct 'haze layer', viewed from slightly above and 'downsun'.

Air-to-ground visibility above a haze layer may be reasonable and can often be improved by flying higher above the haze.

However, visibility within the haze layer may be markedly worse if the slant (oblique) range through the haze layer is increased. It is quite possible to arrive overhead an airfield and be able to see it clearly when looking vertically down through the haze layer. However, as the aircraft descends into the haze, the pilot has to look obliquely through a much greater thickness of the haze layer and may well lose sight of the airfield.

Figure M6.17
Poor in-flight visibility in a haze layer.

Figure M6.18
In-flight visibility when looking into sun in haze can be almost non-existent.

Visibility in haze tends to be much worse when looking into the sun. This can be a particular problem if, for instance, approaching a westerly runway in the late afternoon when the glare of the low sun ahead and the haze combine to reduce visibility significantly. In such conditions, visibility 'down-sun' is likely to be markedly better. Conversely, at night visibility is often better looking towards a low moon.

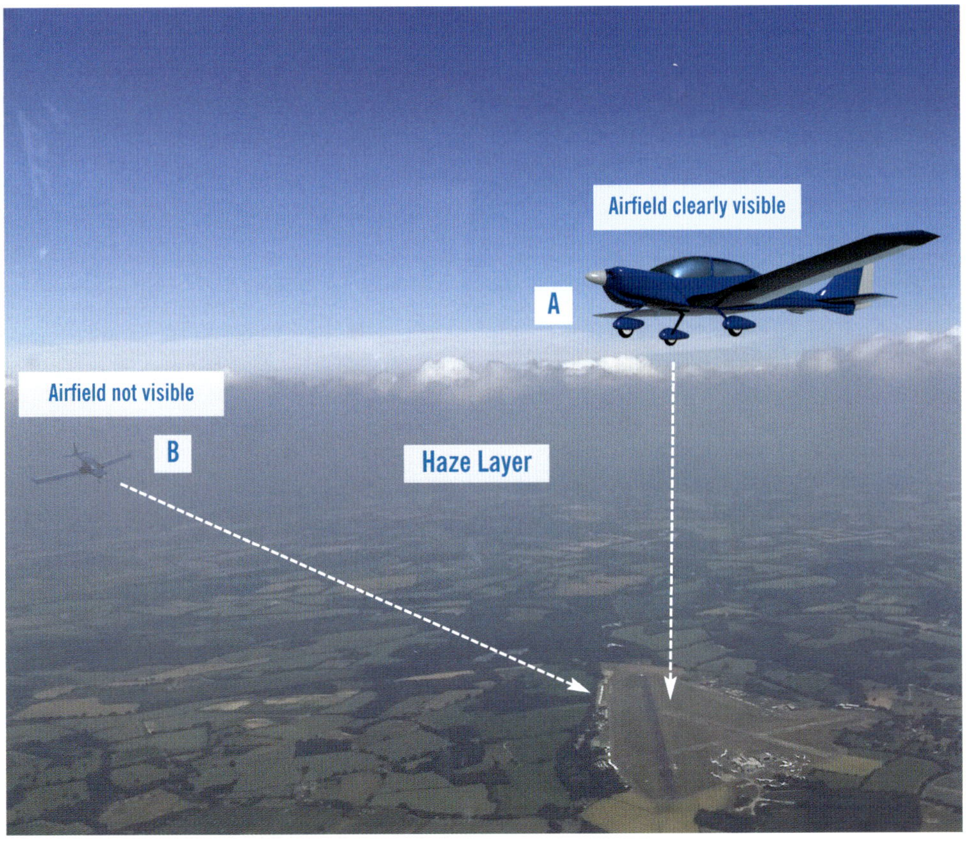

Figure M6.19
Directly overhead the airfield, at position 'A' the pilot can see straight down through a shallow layer of haze, and the airfield is clearly visible. At position 'B' the pilot is looking obliquely through a much greater thickness of haze and cannot see the airfield.

Figure M6.20
Flying lower within a haze layer may give marginally better visibility of the surface, but at the risk of a close encounter with obstructions.

Figure M6.21
When clouds form at the top of the haze layer, visibility beneath those clouds is often significantly worse than away from the clouds.

Figure M6.22
An extract from a METAR observation at Gibraltar airport (ICAO location indicator LXGB). On the 26th at a time of 0650Z (UTC), the visibility is 8000 metres in haze.

METAR: LXGB 260650Z 31002KT 8000 HZ

Within haze layers there may also be quite pronounced local variations in visibility. Visibility near the top of a haze layer is often worse than visibility lower down. Where cumulus clouds have formed at the top of the haze layer, the visibility directly under these clouds can be significantly worse than elsewhere – due in part to the increased humidity. In these conditions it is better to avoid such areas, either by flying lower if it is safe to do so, or by avoiding flying close to the cloud bases.

Once haze has formed, it tends to persist and gradually worsen until a weather system such as an active depression (low pressure area) moves through and clears the atmosphere.

Smoke (FU) Smoke from industrial, domestic or natural fires reduces visibility downwind of the source. In general terms, light winds mean that smoke does not disperse easily; an inversion will also prevent the dispersal of smoke. The reduction in visibility is usually worse downwind of concentrated industrial areas, and during winter anticyclonic conditions.

Dust (DU) and Sand (SA) To a meteorologist, these two phenomena are differentiated by the particle size (dust particles are smaller) and are not confined to desert areas. Dust can reach Europe in the summer, when the airflow is directly from north Africa. Visibility can be significantly reduced; if it rains, dust particles can be 'washed out' in the raindrops causing 'coloured rain'.

Blowing dust or sand can occasionally affect an airfield during strong wind conditions, especially in the case of a coastal airfield near extensive sand dunes. Coastal areas can also be affected by sea spray, which reduces visibility in strong winds. However, since the wind needs to be at least severe-gale strength for this to happen (Force 9, which implies 41 to 47 knots mean speed and gusts up to 60 knots), reduced visibility is likely to be just one of the pilot's major concerns at the time!

Visibility in Precipitation

Any precipitation will reduce visibility to some extent. Slight or moderate rain tends not to reduce visibility greatly, but a very heavy downpour, such as that associated with a thunderstorm, may reduce visibility down to 1000m or less. Drizzle is associated with low cloud and high relative humidity. As such, drizzle nearly always means visibility of less than 3000m.

Hail will cause a marked reduction in visibility. However, it is a hazard in its own right as hail can cause significant airframe damage and it is associated with thunderstorms, so a proficient VFR pilot will avoid flying in such precipitation anyway.

Figure M6.23
Even moderate snow will quickly reduce visibility to 1000m or less.

EGYM MARHAM

METAR EGYM 181120Z 08016KT 0500 +SHSN...

Figure M6.24
An extract from a METAR observation at Marham airport (ICAO location indicator EGYM). On the 18th at 1120Z (UTC) the visibility is 500 metres in heavy showers of snow.

Falling snow has a dramatic effect on visibility. Even light snow is likely to reduce visibility to 6 kilometres or less and a moderate snow shower can reduce visibility to below 1000m; flying into falling snow can mean that all visual reference will quickly be lost. In heavy snow, visibilities of less than 300m are quite common. Even on the ground, a pilot could get lost at their home airfield! Here again, flight into falling snow is not something to be undertaken lightly by a VFR pilot.

Progress Check

38. What is indicated by the meteorological code ‘BR’
39. How may advection fog form?
40. What conditions are most likely to lead to radiation fog?
41. If a daytime airfield weather observation reports visibility as ‘HZ’, in what direction is visibility likely to be the worst?
42. What type of precipitation is likely to lead to the greatest reduction in visibility?

These questions are intended to test knowledge and reinforce some of the key learning points from this section. In answering these questions, a ‘pass rate’ of around 80% should be the target.

Model answers are found at page M174.

M7

Air Masses

Air Masses

Characteristics of Air Masses

Progress Check

M7 Air Masses

Characteristics of Air Masses

An air mass is a large body of air with very similar temperature and humidity over an extensive area, covering maybe millions of square miles. An air mass gains its properties when a large area of slow-moving air remains over a particular region for several days or weeks and once the air mass leaves its 'source' region, meteorologists can predict the type of weather it will bring with it. This will depend on the source region and the original characteristics of the air, the route it has taken since it left its origin, and how quickly it has moved.

An air mass is initially classified by temperature. Around western Europe an air mass can be polar (from the north) or tropical (from the south). It can then be further sub-classified by humidity, depending on whether it formed over the sea (maritime) or over land (continental). Hence there are four principal types of air mass associated with the weather in western Europe.

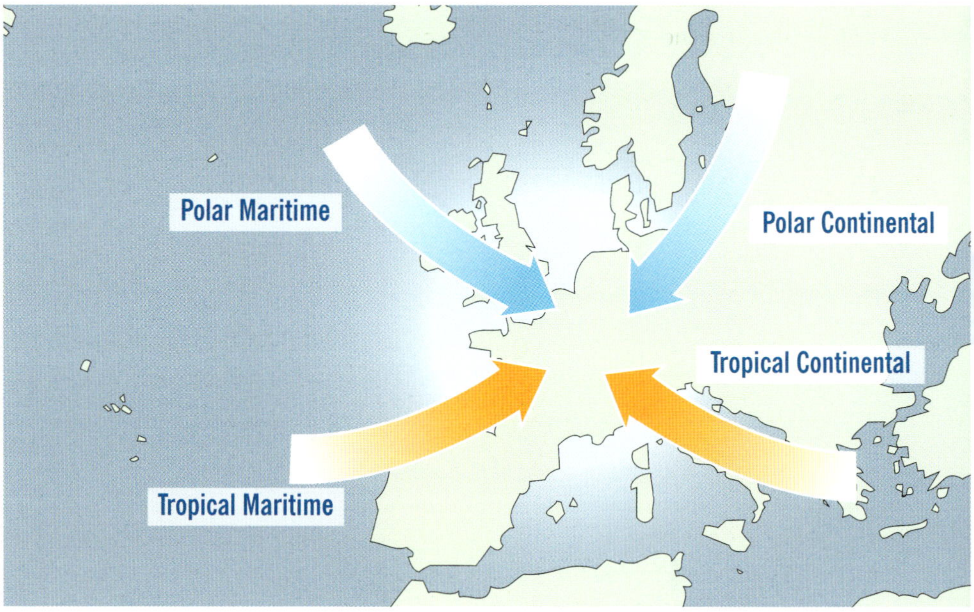

Figure M7.1
The principal air masses which reach Europe.

A **tropical maritime air mass** commonly affects northern Europe in the prevailing south-westerly winds of this region. Tropical maritime air originates in a sub-tropical anticyclone which tends to form and centre over the Azores (for which reason it is sometimes known as the 'Azores High'). This air mass is predominantly warm and humid, and usually brings a stable air flow on reaching Europe because it is cooled from below on its journey northwards. In winter, a tropical maritime air mass tends to bring dull weather with low stratus cloud, poor visibility and even advection fog (especially along south-west facing coasts). During the summer, surface heating effects cause cloud to dissipate inland and visibility to improve, although radiation fog becomes a possibility if the other necessary conditions prevail.

A **tropical continental air mass** comes from the south and south-east, having originated over north Africa and southern Europe. It is hot and dry and, because it is cooled from below on its northerly path, it tends to be stable. This type of air mass brings 'heatwave' weather in the summer with almost cloudless skies and high temperatures. However, visibility is often poor due to haze and dust.

A **polar continental air mass** usually arrives from Russia or Scandinavia in the winter. It is essentially cold and dry and usually brings clear skies and excellent visibility – although visibility can be reduced if the air mass has picked up smoke and dust when crossing heavily industrialised areas. If polar continental air crosses a sea during winter, it can pick up enough moisture from the relatively warm sea to trigger off convective clouds (eg cumulus) and showers on east-facing coasts.

A **polar maritime air mass** originates over the Atlantic near the Canadian Arctic or Greenland. It begins life as relatively humid, cool and initially quite stable air. However, on its passage southwards over the Atlantic, this air mass is heated from below by the warmer seas, so the moisture content of the air mass increases and it tends to become unstable. Arriving in Europe, polar maritime air brings convective (cumulus-type) clouds which are often well developed and can produce showers or even thunderstorms. Outside these showers, visibility tends to be very good.

Polar maritime air may also reach Europe by a less direct route. A slow moving depression in the mid-Atlantic can route the polar maritime air mass well to the south before it approaches Europe from the south-west. Such an air mass is heated from below on its southerly passage, then cooled from below as it heads northwards again. It follows that the lower layers tend to be stable, although unstable air may persist higher up. This so-called 'returning polar maritime' air mass often brings low cloud to south-west coastlines, but extensive heavy showers and thunderstorms further inland, especially in the summer.

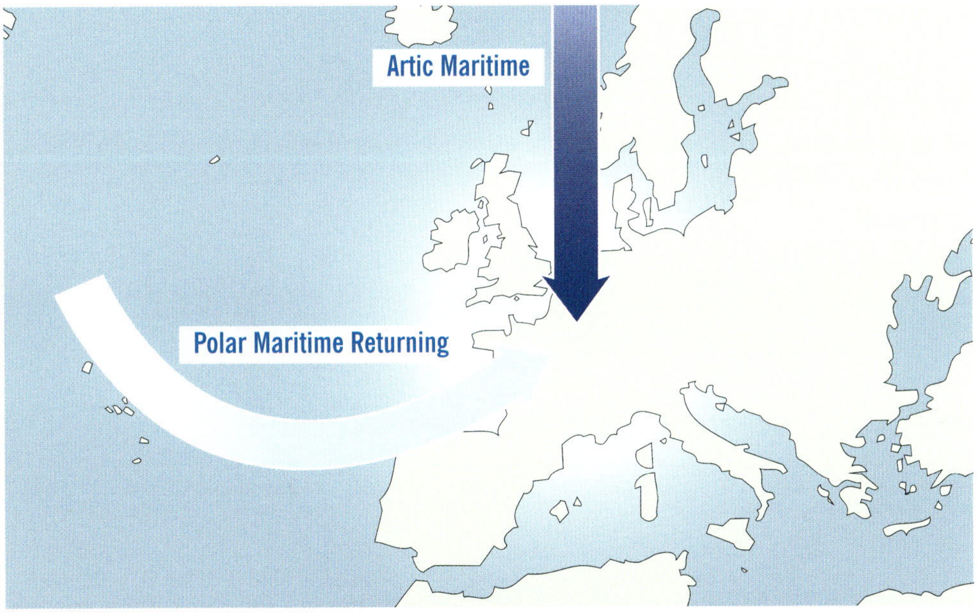

Figure M7.2
Less common air masses reaching Europe.

Figure M7.3
A congested sky with several closely packed Cumulonimbus (CB) clouds – typical of a Polar Maritime Returning air mass.

An air mass which occasionally reaches Europe directly from the north is the **arctic maritime air mass**, whose source region is over the Arctic ocean. Such air is cold and fairly dry and it will bring excellent visibility in the clean, unpolluted air. Any heat and moisture picked up en route may give rise to snow showers over north facing coasts. This is predominantly a winter air flow, which tends to bring the first snowfall of the year to northern Europe and can lead to very low night-time temperatures inland.

Progress Check

43. What are the general characteristics of a Tropical Maritime airmass reaching western Europe in the winter?
44. What is the name of the air mass which originates in the north Atlantic near Newfoundland and Greenland, and usually brings an unstable air flow from the north west?

These questions are intended to test knowledge and reinforce some of the key learning points from this section. In answering these questions, a 'pass rate' of around 80% should be the target.

Model answers are found at page M175.

M8 Low Pressure Systems

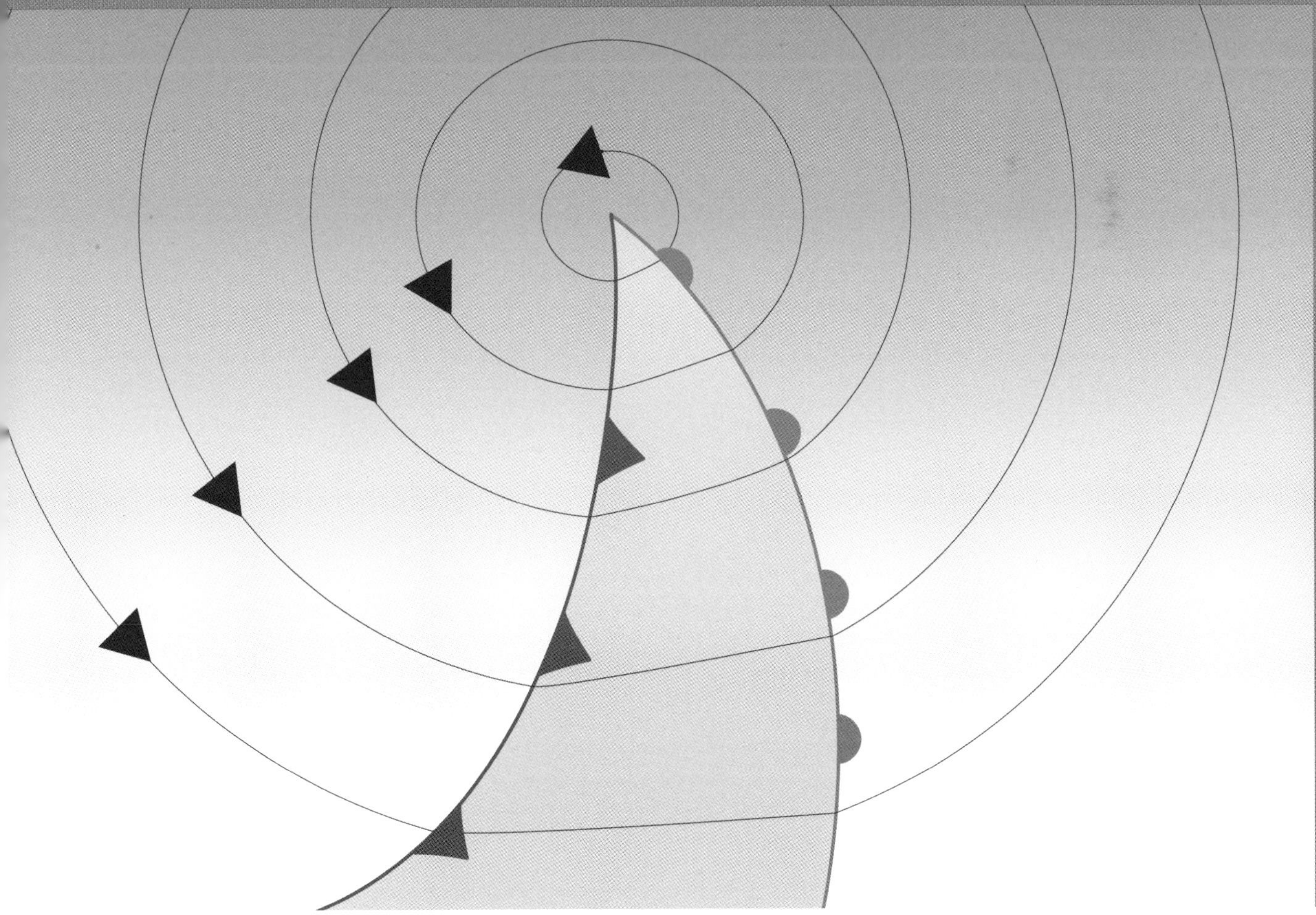

M8 Low Pressure Systems

The 'Warm Sector' Depression

Meteorology is at best an inexact science because, in the mathematical sense, weather systems are 'chaotic'. No two depressions are exactly alike, nor any other two meteorological phenomena. The purpose of attempting to describe a 'standard' warm-sector depression is to provide basic guidelines, which can be used to assess real-life situations. What follows should be taken as general guidelines, not everything described below is absolutely applicable to every depression.

Most depressions that reach western Europe are born over the North Atlantic, along the **polar front**.

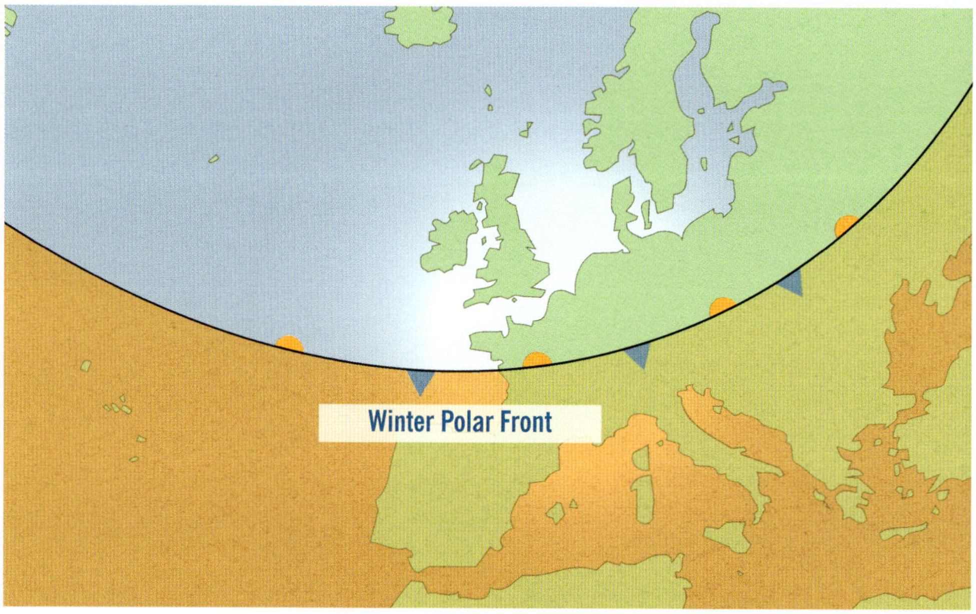

Figure M8.1
The Polar Front marks the boundary between cooler polar air to the north, and warmer tropical air to the south. In winter, the front typically sits across central Europe, steering depressions into the middle of Europe.

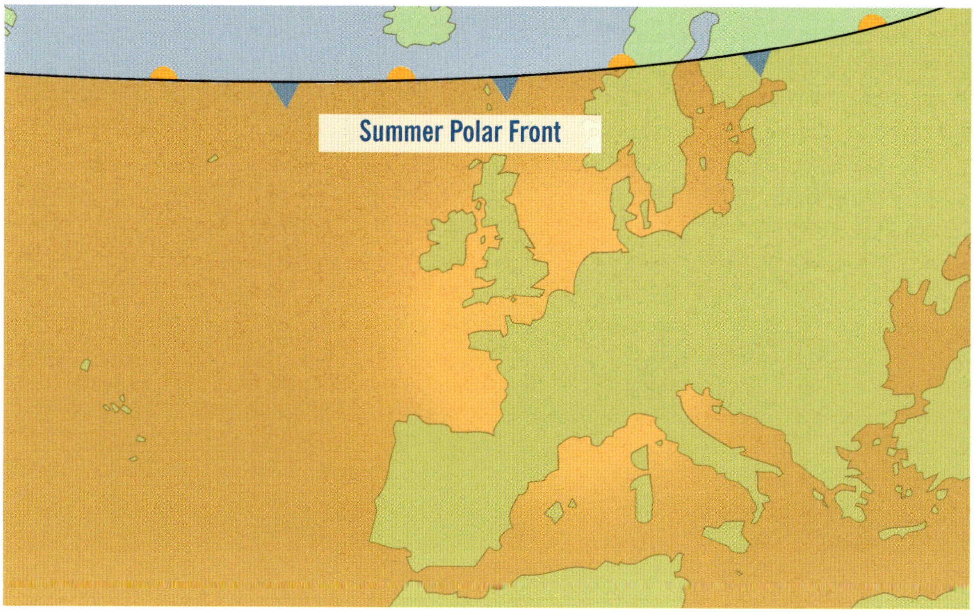

Figure M8.2
In summer, the Polar Front typically sits mostly to the north of Europe, steering depressions away from all but the most northern European states.

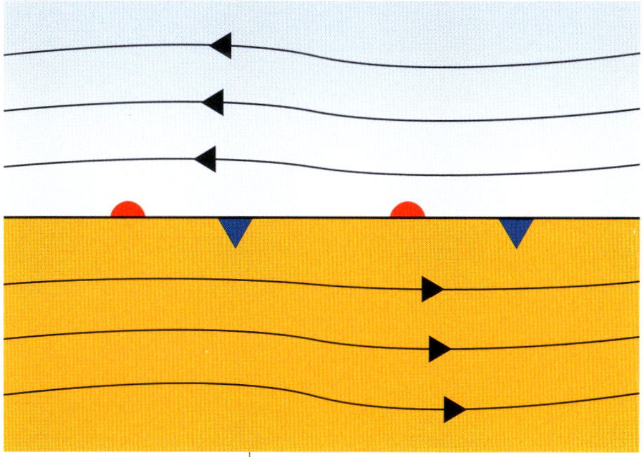

Figure M8.3a
The polar front with cold air to the north and warmer air to the south. The air is flowing in different directions either side of the front.

A **front** is a defined boundary between differing air masses and the North Atlantic polar front usually lies approximately east/west, marking the boundary between cold polar air to the north, and warm moist air to the south. The position of the polar front is not fixed and it often changes on a daily basis; in very general terms it lies approximately along the 60°N parallel of latitude in the summer (taking depressions into the most northerly European regions) and along the 50°N parallel in winter (bringing depressions into mid-Europe).

A depression begins as an unstable wave along the polar front. The fine-scale mechanisms involved in the formation of depressions are complex, it's enough to say that they are associated with the **jetstreams** flowing above the polar front. Jetstreams are fast-moving streams of air at high altitude that tend to form over transition zones such as the polar front. The initial formation and the subsequent movement of the depression are closely linked with the location and path of the associated jetstream.

Figure M8.3b
At the surface, a 'bulge' of warm tropical air pushes into and rises over the colder, more dense polar air.

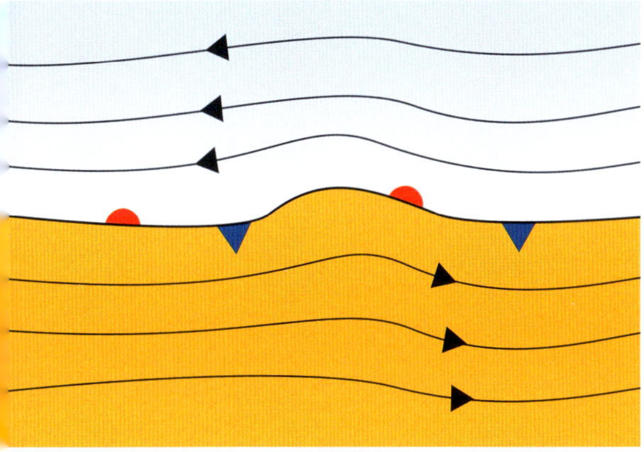

Figure M8.3c
As the warm air rises in the 'bulge' and begins to rotate anti-clockwise, the pressure at the centre begins to fall.

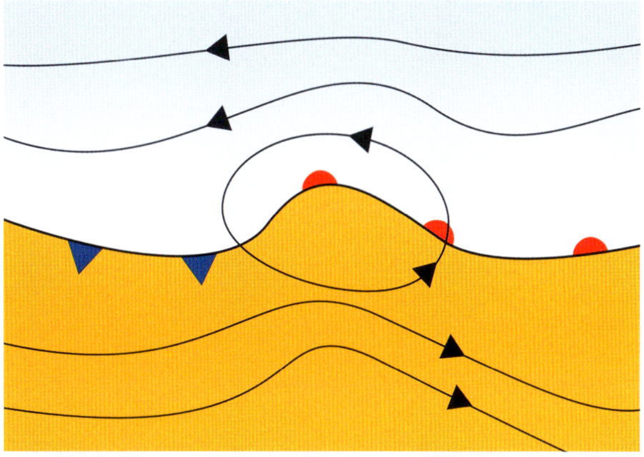

At the surface the first sign of a depression forming is a 'bulge' of warm air into the colder air. Warm air rises at this point and the pressure at the surface begins to fall. As more air flows (converges) into this area it acquires a curve in its flow. The combination of curvature and ascent of the air flow causes the low to begin to 'wind up'. The process is cumulative, and the consequence is an increasing mass of rising air. This rising air cools and condenses to form cloud, and the energy released from this condensation process strengthens the depression even more. As the centre of the low deepens (the atmospheric pressure reduces further) the air flow becomes circular. Soon there are well-defined boundaries between the differing air masses. The area of warmer tropical maritime air between the fronts is called the **warm sector** – which gives its name to this type of depression. The air outside the warm sector is polar maritime air, occasionally referred to as the 'cold sector'. By now the depression is moving eastwards, rotating anti-clockwise, generally following the path of the jetstream and gathering energy as it continues to deepen (the atmospheric pressure at its centre still falling).

Figure M8.3d
As pressure at the centre continues to fall (the depression is 'deepening'), the whole depression begins to rotate anticlockwise. There are now well-defined boundaries between the air masses.

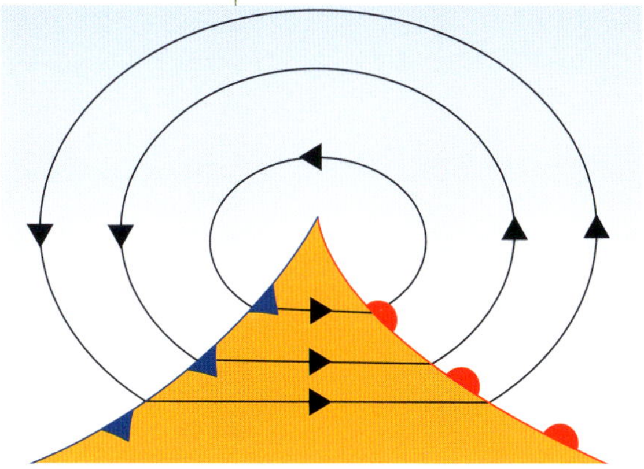

The Warm Front

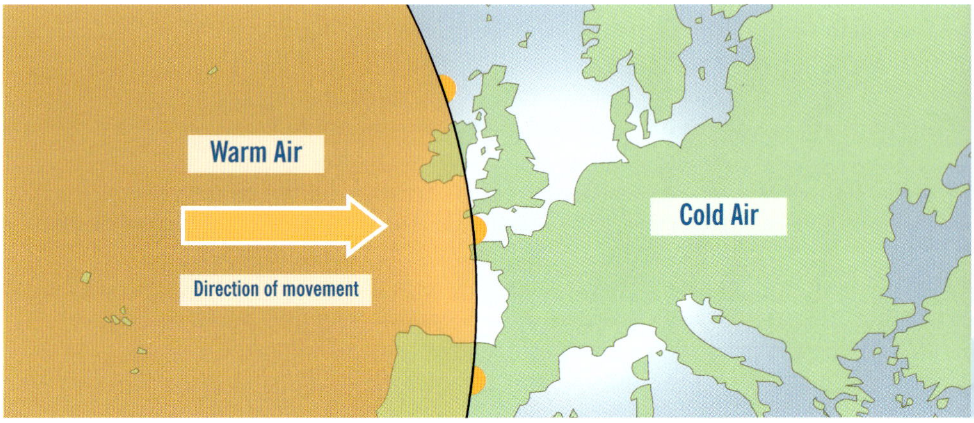

Figure M8.4
A vertical view of a warm front – bringing warm air to replace colder air.

In meteorology, a 'front' is a boundary between two air masses and is primarily defined by air temperature. If a front brings in a warm air mass, then logically enough it is called a **warm front**.

In a classic warm-sector depression, the warm front is formed where warm (less dense) air replaces colder (more dense) air. Typically, it is the warm front which will bring the first visible signs of an approaching warm sector depression.

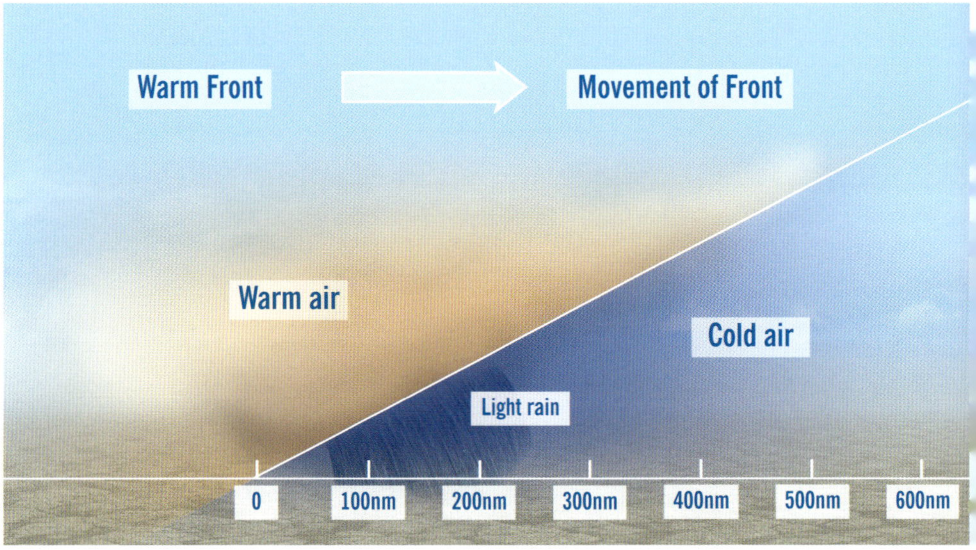

Figure M8.5
A cross section through a 'typical' warm front – with gently sloping frontal surface, increasing amounts of lowering layered cloud, a wide band of precipitation and a wide transition zone into the warmer air behind the front.

The slope of the warm front is generally quite gentle, the commonly-quoted figure is a slope gradient of about 1:150. At the top of the frontal slope, close to the tropopause, cirrus-type clouds are found. These can be up to 500nm (925km) ahead of the surface front. As the warm front approaches, the cloud (usually stable and layer-type) thickens and lowers. Rain or snow first falls from altostratus cloud at around 15,000ft, but tends to evaporate before reaching the ground. Slight precipitation first reaches the surface 100-200nm (185-370km) ahead of the surface warm front. This precipitation increases to moderate and becomes continuous as the warm front approaches. All this time the pressure is steadily falling as the warm front nears. The surface warm front itself is usually marked by the lowest cloud (often nimbostratus), which may come right down to the surface. The transition as the warm front passes is rarely sharp, and the frontal zone of a warm front can be up to 90nm/150km across, which makes for a gradual transition between air masses.

Here is a good place to re-emphasis that this description is a generalisation, a 'typical' model which is unlikely to be exactly replicated in real-life. About the only thing which can be said with certainty about a warm front is its defining feature – the temperature difference. A warm front always brings warmer air to replace colder air.

The first sign of the passage of a warm front within a depression may well be a veering of the surface wind, followed by a steadying of the pressure (which has usually been falling ahead of the front) and an increase in both temperature and dew-point. The precipitation should peter out after a while, although low cloud, drizzle and poor visibility often persist behind the front in the warm sector (which in Europe is usually tropical maritime air).

The Cold Front

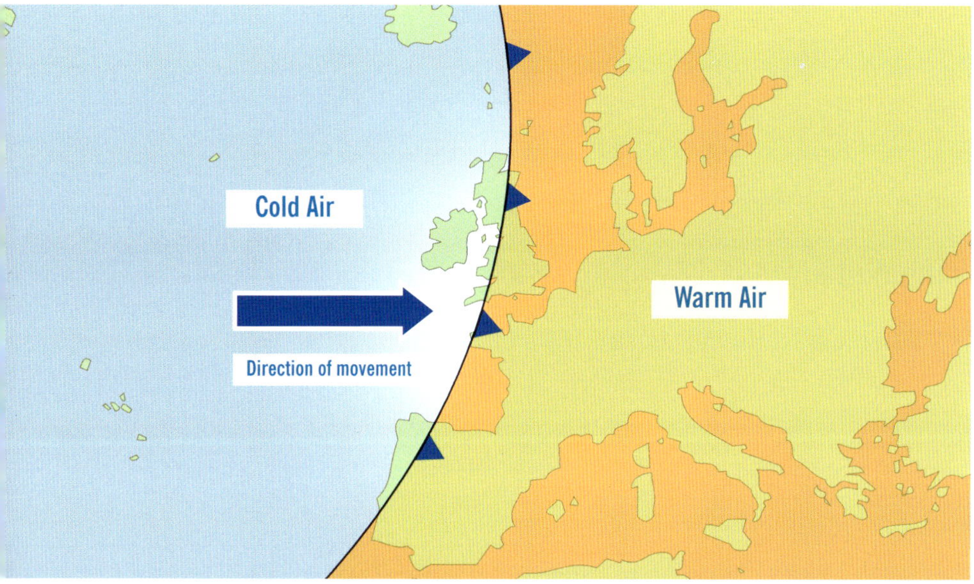

Figure M8.6
A vertical view of a cold front – bringing colder air to replace warmer air.

By definition, a **cold front** marks the boundary of a cold air mass replacing warmer air. In a typical warm-sector depression, the cold front marks the leading edge of a colder air mass replacing the warmer, less dense, air of the warm sector.

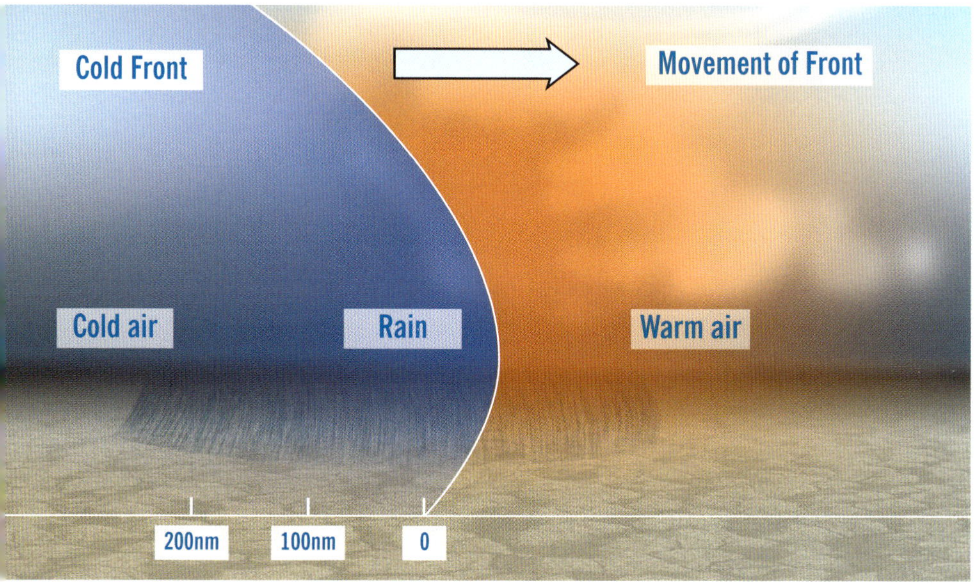

Figure M8.7
A cross section through a 'typical' cold front – with relatively steep frontal surface, extensive vertical cloud at the front, a narrow band of precipitation and a relatively narrow transition zone into the colder air behind the front.

The slope of the cold front tends to be steeper than that of the warm front, the 'average' cold front having a slope of around 1:50. It follows that the zone of cloud and rain along the cold front tends to be narrower and more concentrated than the warm front and is more likely to include convective cumulus-type cloud of significant vertical extent, possibly producing intense showers of heavy precipitation. Pressure falls ahead of the front.

The actual passage of the surface cold front may be marked by especially heavy precipitation and 'squally' conditions with strong, gusty winds. An active cold front can be marked by a line of cumulus or cumulonimbus clouds, although these may be hidden to the ground-based observer by lower and more layered clouds persisting in the warm sector.

The passage of a cold front is often marked by a dramatic change in weather. The cloud and rain often disperse quickly, leaving drier and much clearer conditions behind the cold front. The temperature and dew-point fall but the pressure rises. The surface wind often veers sharply and becomes 'gusty' with the passage of a cold front, as a polar maritime air mass is introduced.

Sometimes a slow-moving and fairly inactive depression leaves a long cold front trailing back into the Atlantic. A 'secondary' depression can form within the 'old' primary depression somewhere along this trailing front. Secondary depressions can then deepen quickly, bringing a rapid deterioration in the weather where previously the meteorological situation had become quite settled.

Occluded Fronts

In a 'classic' warm sector depression, the cold front moves faster than the warm front. Eventually, the cold front overtakes the warm front and when this happens, the air of the 'warm sector' between the warm and cold fronts is lifted clear of the surface, usually as the faster moving cold front 'undercuts' the warm front. Although the air either side of the **occluded front** at the surface belongs to the 'cold sector', there will inevitably be some temperature difference across the front, albeit less distinct than that for a warm or cold front on its own.

Figure M8.8
An idealised chart of a maturing warm sector depression showing the well-developed warm and cold fronts.

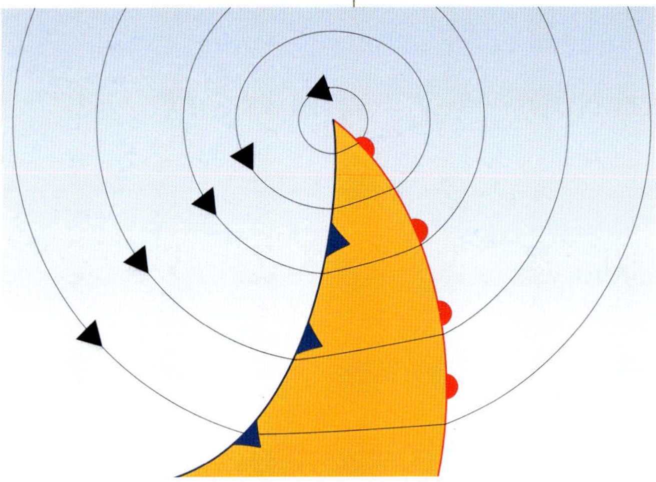

Figure M8.9
As the cold front overtakes the warm front in the depression, an occluded front is formed, which can bring extensive low cloud and precipitation.

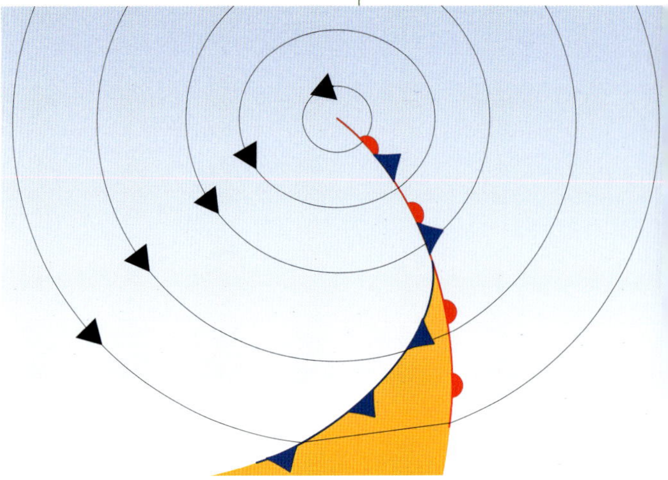

Figure M8.10
Cross-sections through a 'cold' occluded front (left) and a 'warm' occluded front (right). Either sort is likely to bring extensive cloud and prolonged precipitation.

Where the cold front has undercut the slower moving warm front, the air behind the surface front is colder than that ahead, this is known to meteorologists as a 'cold' occlusion.

Less commonly, the cold front rides up over the slope of the warm front, leaving less cool air behind the surface front than ahead, in which case the result is a 'warm' occlusion.

In both 'warm' and 'cold' occlusions the structure of an occluded front tends to be complex and difficult to generalise and in any case, the differentiation between warm and cold occlusions is of more interest to meteorologists than pilots. Typically, an occluded front has all the cloud and precipitation of the warm and cold fronts together – and a slow-moving occluded front can bring persistent low cloud and heavy (and prolonged) precipitation. Often, but not always, there is little change of wind velocity across the occlusion.

The 'occluding' of a depression marks the beginning of the end of its life-cycle. The low pressure area gradually fills (the pressure at the centre rises) and it becomes slower moving. The fronts become less active and the winds decrease. Meanwhile, back in the North Atlantic at the polar front, a new depression may well be forming.

The fronts that exist within a 'typical' warm sector depression crossing Europe tend to travel across the surface at an average speed of 20 – 30 knots. However, a front may move much more slowly than this, or even stop, in which case it becomes known as a 'quasi-stationary' front. Because such a front is going nowhere, the conditions it may be bringing (eg low cloud, poor visibility and precipitation) are obviously going to persist for some time.

Figure M8.11
A 'quasi-stationary' front.

It is not uncommon for a 'quasi-stationary' front to have 'waves' moving long it – similar to the wave pattern formed on a flag fluttering in a breeze. On a weather chart a 'quasi-stationary' front will be shown with alternating cold and warm front symbols on either side of the front because as waves move along the front one side of the wave is bringing warm air to replace cold air and vice versa.

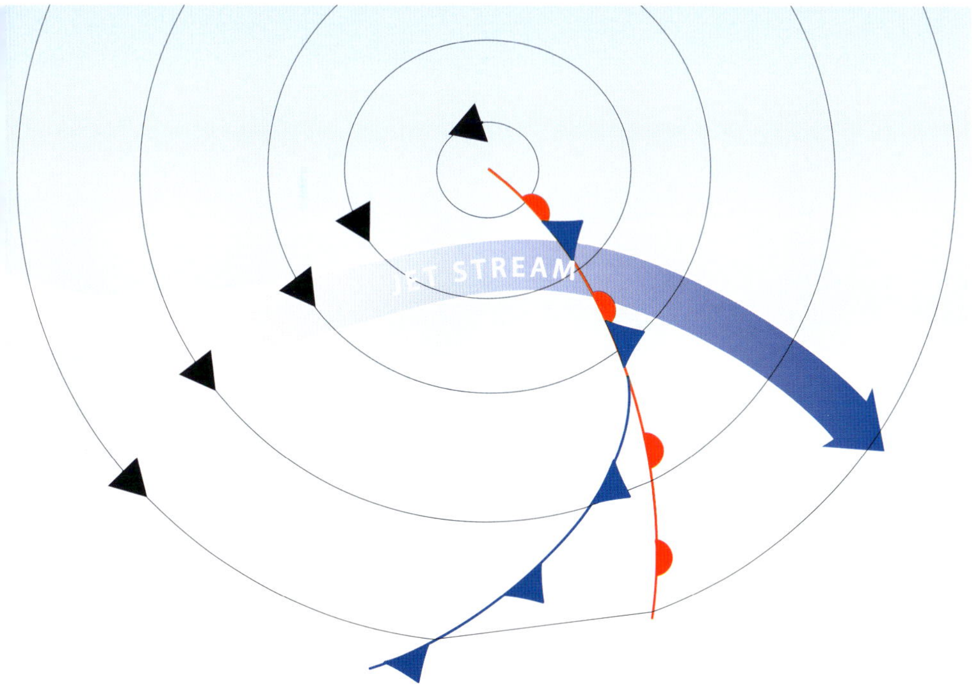

Figure M8.12
The development and track of a warm sector depression is closely related to the upper winds of the jetstream.

In meteorology it is difficult and sometimes dangerous to try to lay down hard and fast rules about the type of conditions that certain atmospheric conditions will cause. A front can bring extensive cloud and rain and a sharp veer in the wind, or it may pass almost unnoticed. It is sometimes thought that *all* warm fronts bring extensive stratus-type cloud and *all* cold fronts bring big cumulus build-ups. This is simply not true. A warm front can (rarely) have embedded cumulonimbus clouds associated with it, giving thunderstorms. Equally, a cold front can have extensive layered stratus cloud. However, there are some generalisations about warm-sector depressions which *tend* to be valid:

- The track of the depression is largely guided by the jetstream. From surface observations the direction and speed at which the depression is moving is indicated by the isobars in the warm sector.
- Travelling depressions tend to steer around established anticyclones.
- Once a depression has occluded and begun to fill up (ie the pressure at the centre is rising) the fronts tend to become less active. A depression that is still deepening (meaning that the pressure at the centre is falling) brings more active fronts and generally worse weather.
- When a string of depressions is passing through Europe, there is often a ridge of high pressure between each one. This gives a brief period of fine weather; unfortunately, this may occur overnight!
- The weather at the fronts is usually worse near the centre of the depression and better away from the low, or where the fronts cross an area of high pressure.
- The greater the temperature difference and the steeper the temperature gradient across a front, the more active and sharply defined the front will be.
- At the surface, the change of temperature with the passage of a front is not always well-defined. Cloud and rain at the passage of a warm front may actually reduce surface temperature: sunshine in clear skies behind a cold front may briefly increase it. Veering of the wind and change in the dew-point are usually more reliable indicators of the passage of a front.
- The sharper the wind veer associated with a front (as seen by a sharp 'V' in the isobars at the front), the more active the front will be. An acute 'V' in the isobars implies a strong wind shear as the wind rapidly veers with the passage of the front. Active cold fronts are particularly good at producing wind shear.
- Fronts may be weak (inactive) or strong (active). In general, a strong warm front is followed by a weak cold front and *vice versa*.
- There is often an area of very cold, clear air just behind a cold front (a 'dry slot') which may bring clear skies for a few hours until cumulus clouds and showers move in.

Figure M8.13
The approximate areas of low cloud and precipitation in a 'classic' warm-section depression.

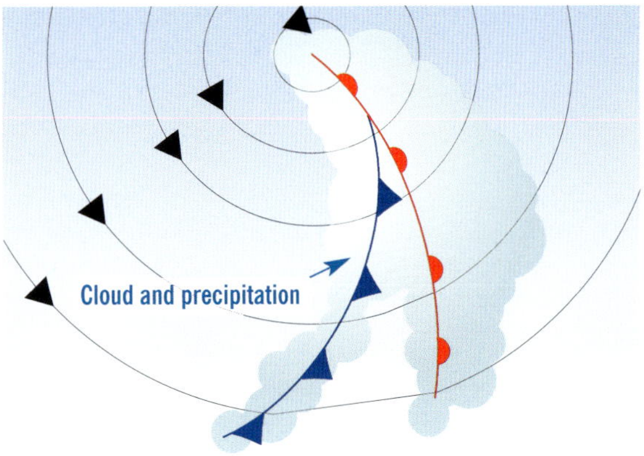

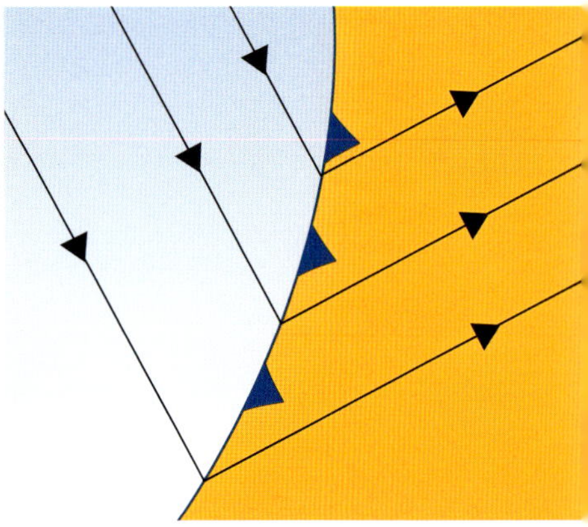

Figure M8.14
A sharp veer of wind direction is possible with the passage of a front. The more acute the 'V' in the isobars across the front, the more marked the veering of the wind will be.

Troughs and Convergence

When differing horizontal air flows converge, the result is that air is forced to rise, with a consequent reduction in surface pressure. Such **convergence**, with attendant rising air and reducing pressure, takes place on a large scale within a developing depression and is responsible for air rising and cooling over a wide area, often leading to the formation of cloud and possibly precipitation.

More localised convergence takes place at the fronts, tending to increase the cloud, precipitation and other features of the front for as long as the depression continues to 'deepen' (that is, the pressure at centre of the low is decreasing).

Convergence can also form an elongated area of low pressure known as a **trough**. A trough often brings weather similar to that associated with a front, but without the change in air mass. Such organised trough lines may be marked on synoptic weather charts and are often found in well-developed lows (depressions). In western Europe, troughs are often (but not always) associated with unstable air and convective (eg cumulus or cumulonimbus) clouds.

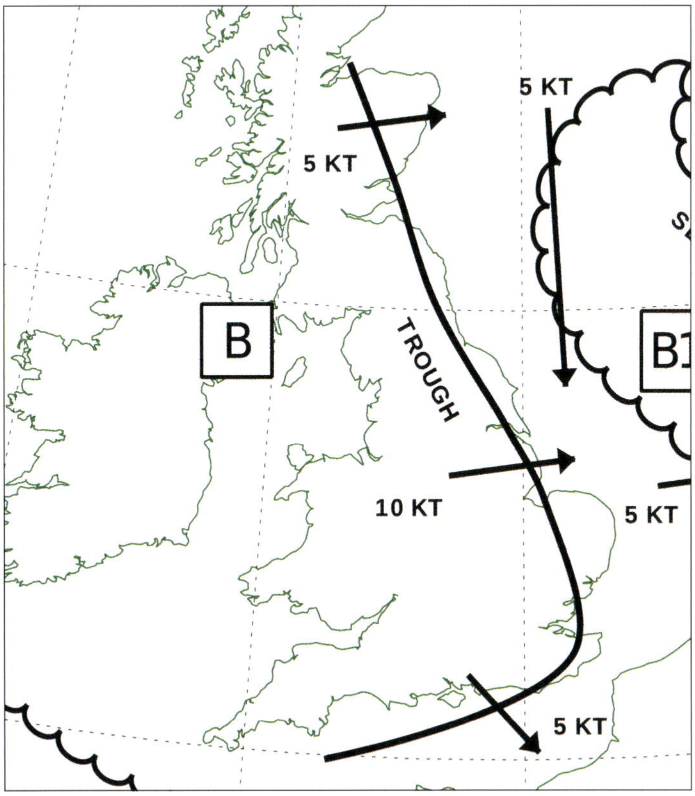

Figure M8.15
A 'trough', shown here as a solid black line, can bring weather similar to a front, but without the change in air mass.

Progress Check

45. What is the definition of a 'front'?
46. What is the general relationship between the movement of a depression and the jetstream?
47. What is the name of a front which brings warm air to replace cooler air?
48. List the weather conditions likely to occur after the passage of a cold front
49. What name is given to an elongated area of low pressure?

These questions are intended to test knowledge and reinforce some of the key learning points from this section. In answering these questions, a 'pass rate' of around 80% should be the target.

Model answers are found at page M175.

M9 High Pressure Systems

M9 High Pressure Systems

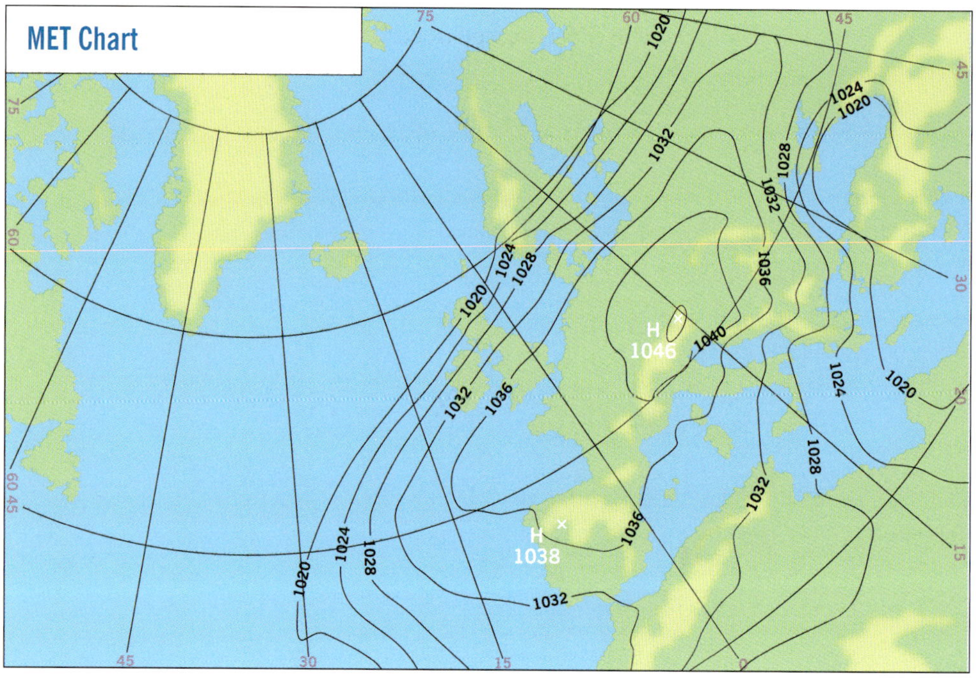

Figure M9.1

A high pressure anticyclone over central Europe.

Anticyclones

An **anticyclone** is an area of high pressure, featuring the widespread descent and consequent **divergence** of air. The air in an anticyclone sinks (or 'subsides') slowly, an average rate being about 2000ft per day. As air in an anticyclone subsides, it is compressed and so it warms. If the air is dry, it will be warming at about 3°C per 1000ft as it sinks and compresses.

Figure M9.2

Within a high pressure area air 'sinks' or 'subsides' towards the surface, warming as it does, spreading out and circulating clockwise (in the northern hemisphere).

The general movement of air within an anticyclone is that it sinks to somewhere near the surface and then diverges (spreads out) from the centre of the high, with the wind flowing in a clockwise direction (in the northern hemisphere). Because the pressure gradient within an anticyclone is generally 'slack', with widely spaced isobars, light winds are often a feature of areas of high pressure. Likewise, the general subsidence of the atmosphere in an anticyclone discourages the formation of cloud – although local heating of the surface, especially in summer, may cause small cumulus clouds to pop up. Because the atmosphere in an anticyclone is usually stable, these cumulus clouds do not normally grow to any great extent; they are often called 'fair-weather cumulus', often shortened to 'cu' in conversation. Anticyclones tend to move very slowly, or become stationary, and so the conditions they bring will tend to persist for days or weeks at a time.

Figure M9.3
Typically isolated and insignificant 'fair weather cumulus' in a summer anticyclone.

A summer anticyclone in Europe often brings 'heatwave' weather since its clockwise flow tends to bring tropical continental air from the south, and these conditions can last for long periods as already explained. Consequently, the arrival of anticyclonic conditions is initially good news for pilots with the probability of light winds, clear skies and generally calm and settled flying conditions.

The subsidence within an anticyclone can encourage a **temperature inversion** to form. To recap, a temperature inversion is a layer of the atmosphere through which the temperature warms with increasing altitude – in other words the reverse of the normal situation whereby air cools with increasing altitude.

Level	Temperature
6000ft	+17°C
5000ft	+20°C
4000ft	+23°C Top of inversion layer
3000ft	+23°C
2000ft	+22°C
1000ft	+21°C
Surface	+20°C

Figure M9.4

An inversion acts as a 'lid' on the atmosphere. On a large scale the tropopause itself can be thought of as an inversion, keeping virtually all water content and weather in the troposphere. On a smaller scale, an inversion acts in a similar way. Any rising thermals from the ground stop on reaching the top of an inversion, and wind flow above and below the inversion can be markedly different – the inversion in effect separating the air above it from the effect of the friction layer. A common result of this effect is often a much stronger wind above the inversion than below. Inversions can also have profound effects on the behaviour of VHF radio waves.

Figure M9.5
An inversion, capping steam rising from a cooling tower.

As an inversion layer becomes established, so smoke, dust and haze are concentrated within it. There are no strong winds to disperse these pollutants, and little or no precipitation to wash them out of the atmosphere. Over a period of days the visibility below the inversion layer will worsen, and the height of the inversion layer will gradually rise. This situation is likely to persist until either the anticyclone breaks down (often in a burst of thunderstorms and heavy showers in the summer) or a depression moves in, leaving much clearer air behind it.

In winter an anticyclone brings similar features – at first the clear skies and calm conditions can bring very good (if rather cold) flying conditions, although if temperatures fall below freezing, frosty or icy runways and aircraft are possible. After a few days an inversion layer forms and the resulting haze reduces visibility. The formation of a persistent layer of turbulence cloud may well make VFR flight under the inversion layer difficult. In addition, air very near to the surface may be cooled by its proximity to the cold ground, further promoting the formation of an inversion (as air not close to the cold surface may well be warmer). A winter anticyclone will probably persist until a strong and fast-moving depression moves in.

It can be the case that a temperature inversion forms across a very narrow vertical layer of the atmosphere, typically a few thousand feet above the surface. Below this inversion there are vertical convective currents (thermals), with haze commonly concentrated below the inversion, cumulus cloud may well form at the level of the inversion. Above the inversion layer the air is commonly much clearer and smoother. In the summer, the depth of this 'convective' layer usually increases during the day as surface and air temperatures increase.

Figure M9.7
Fog covering the surface under an autumn anti-cyclone.

In spring and autumn, the light winds and (initially) clear skies within an anticyclone can encourage the formation of radiation fog which may persist for several days. Additionally, the formation of a low-level inversion is more likely. The effect of the separation of the wind flow above the inversion from the friction layer below – leading to marked wind shear at the top of the inversion – is particularly noticeable in these circumstances.

Ridges

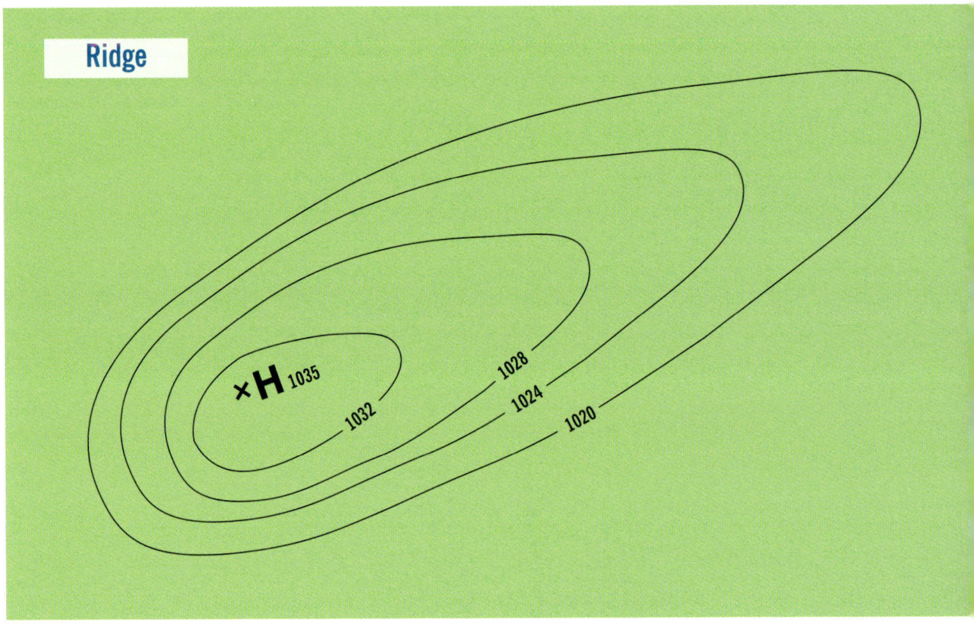

Figure M9.8

A *ridge* is an elongated area of high pressure, usually bringing very similar weather to that of an established anticyclone. 'Weak' ridges are a particular feature of winter weather, when they often appear between successive depressions. They may well bring a short period of fine weather between depressions.

Cols

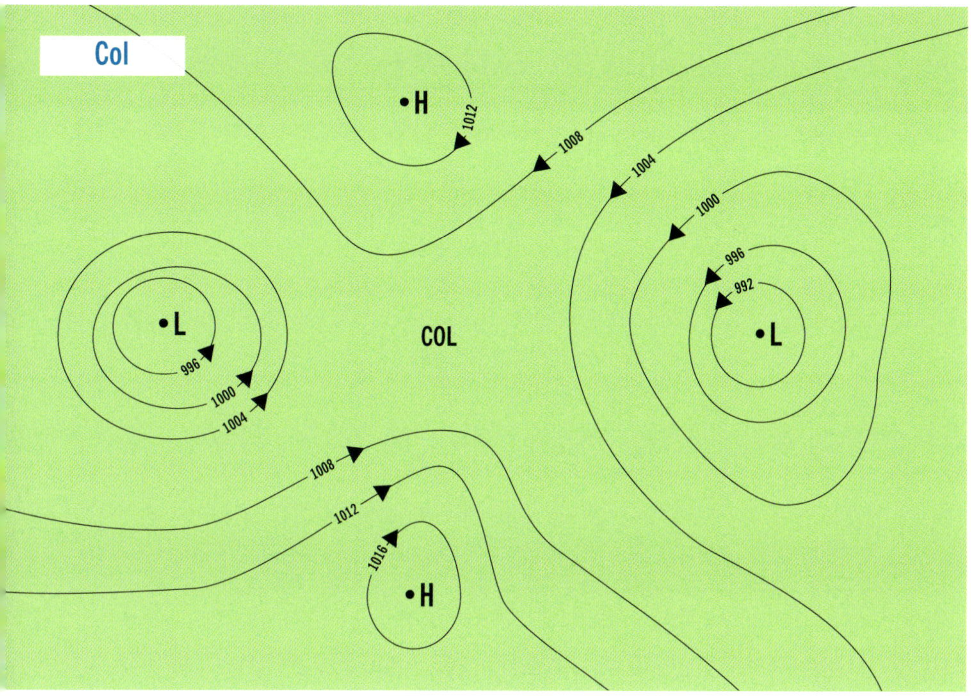

Figure M9.9

A *col* is an area of stagnant air, sandwiched between two highs and two lows. About the only hing that can be said with certainty about the weather within a col is that the winds will be ight (as witnessed by the wide spacing of isobars in a col). The weather situation within a articular col will often be influenced by the most dominant pressure system nearby, but it akes a meteorologist with knowledge of the upper atmosphere to make a sensible forecast. n spring and autumn the light winds can encourage fog: thunderstorms may break out in ummer. But it really is very difficult to make valid generalisations about the weather in and round cols. Each one is an individual case.

Progress Check

50. What is an anticyclone?
51. If temperature increases with height (rather than decreasing), how is this situation described?
52. Which particular weather phenomena may be associated with an anticyclone over land in late Autumn?
53. What is a ridge?
54. What winds should be expected within a col?

These questions are intended to test knowledge and reinforce some of the key learning points from this section. In answering these questions, a 'pass rate' of around 80% should be the target.

Model answers are found at page M175.

M10 Hazardous Weather Conditions: Icing

Airframe icing

Rain Ice

Frost

Piston Engine Icing

Progress Check

M10 Hazardous Weather Conditions: Icing

Airframe icing

At the surface, water rarely exists as a liquid at temperatures below freezing (0°C) because water simply turns into ice once the surface temperature is into minus numbers. Things are different in the atmosphere, however. The water droplets which make up a cloud can remain liquid at temperatures well below freezing: as low as -40°C (although liquid water droplets in air temperatures below -20°C are relatively rare). Water droplets can remain in this liquid, **supercooled**, state until something comes along onto which they can freeze – such as an aircraft flying above the freezing level. So an aircraft flying in cloud above the freezing level is in potential icing conditions.

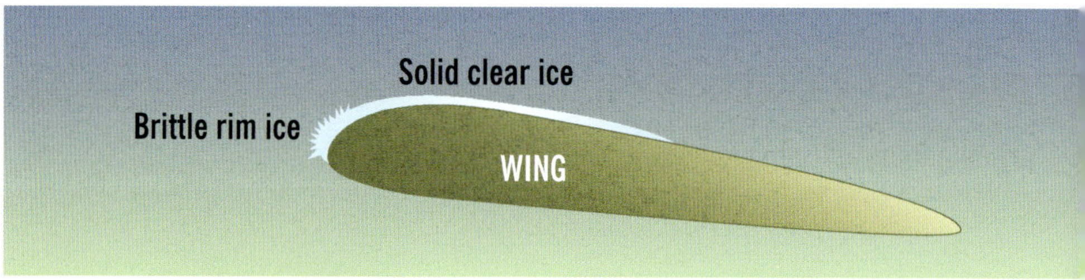

Figure M10.1
When flying through typical icing conditions, brittle 'rime' ice forms on the aircraft leading edges, such as the wing. Denser and more tenacious 'clear' ice forms further back as the supercooled water droplets flow back and freeze.

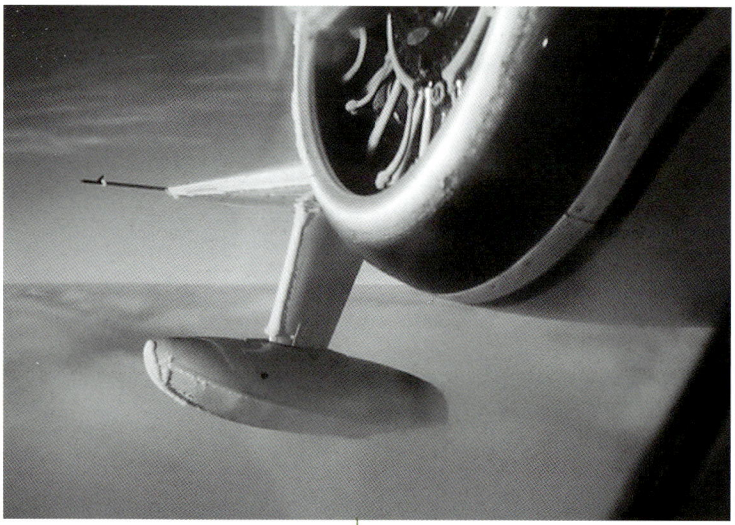

Figure M10.2
Ice on the leading edges of an aircraft which had just flown through a warm front. So much ice accumulated that the aircraft had been forced to descend, even at full power and best climb air speed.

When an aircraft meets a supercooled water droplet, some of the droplet freezes on contact and forms a brittle layer of ice, usually on the leading edges of the wings and tail and also on aerials, struts, etc. This semi-opaque coating is called **rime ice**. A proportion of a larger water droplet may not freeze on initial contact, but only freeze as it flows back across the surface of the aircraft. The sheet of clearer ice formed in this way is called **clear ice** (or sometimes 'glaze' ice).

The worst airframe icing is generally considered to occur at an **Outside Air Temperature** (**OAT**) of between +2°C and -20°C. Icing becomes less severe below -20°C and is fairly rare below -40°C, because at these temperatures clouds mostly consist of ice crystals. However, there are times when these rules of thumb do not apply. Within a cumulonimbus cloud, for instance, severe icing may be encountered at any air temperature below 0°C

Some cloud types give more pronounced icing than others. Cumuliform clouds tend to hold larger water droplets than stratus-type clouds and therefore contain more severe icing conditions. Clouds formed by orographic uplift or by mountain-wave conditions (for example nimbostratus clouds) can also contain worse icing conditions than elsewhere. Stratocumulus cloud can sometimes contain unexpectedly severe icing conditions, particularly just below an inversion over the sea.

As well as flight in cloud, flight in snow can also lead to icing. In general, snow is less likely to accumulate on an aircraft in the way that supercooled water droplets do. However, snow can stick on leading edges and clog air intakes, a process sometimes called 'impact' icing.

Any form of icing is a hazard to an aircraft. Ice on the airframe will markedly increase drag and reduce lift – it may become impossible to maintain level flight even at full power. Likewise, an aircraft's flight characteristics can be so drastically compromised that the aircraft becomes uncontrollable. Engine intake icing can cause a marked reduction in power, or even a total engine failure. Ice can form on the windscreen, block the external pitot and static pressure sources, form inside the lines of the pitot/static system or even freeze the flying controls so that they become immovable. In short, icing conditions and a light aircraft without icing protection is a dangerous combination.

A pilot without some form of instrument flying qualification, and a suitably equipped aircraft, is not permitted to fly in cloud, let alone fly in cloud at temperatures below 0°C. Therefore, a pilot without any instrument qualifications should already be avoiding the conditions that most commonly lead to airframe icing.

It is also true that the vast majority of light aircraft are not fitted with airframe de-icing equipment. Consequently, the Flight Manual for such aeroplanes will usually state emphatically that the aircraft is not cleared for flight into known icing conditions. The potential consequences of ignoring this advice are severe.

Rain Ice

This is an uncommon form of icing that, unusually, can occur outside cloud. The classic environment for **rain ice** is the area just ahead of a warm front in winter. The clouds of the warm front are mostly within the 'warm sector' and so precipitation may fall as rain, through the front, and into the colder air ahead of the front. If this rain meets an object whose surface temperature is below freezing (such as an aircraft) the rain drops may freeze on contact – forming lumps of clear ice which are extremely difficult to remove.

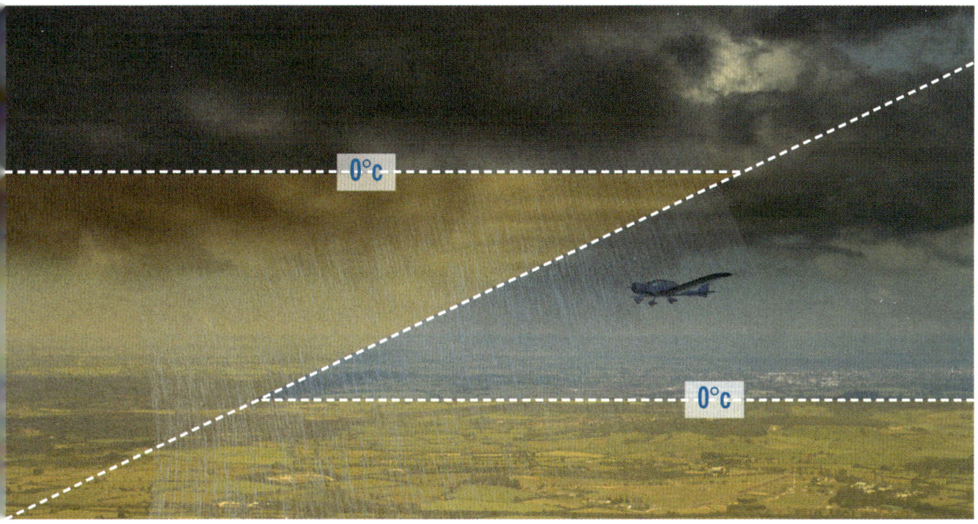

Figure M10.3

The conditions ahead of a warm front in winter are those most likely to lead to rain ice.

Figure M10.4a (far left)

Rain ice on a car left outside in winter. The car park was effectively an ice rink...

Figure M10.5b

...the car windscreen seen from inside, the view outside completely obscured by thick rain ice. Even with the engine running and full 'windscreen clear' mode, it took around 10 minutes for the windscreen to clear.

Figure M10.6
An aviation weather forecast (TAF) for Edinburgh airport predicting the possibility of FReezing RAin (FZRA) between 16:00 and 21:00 (UTC) on the 15th.

EGPH EDINBURGH AIRPORT

TAF EGPH 150501Z 1506/1606 15008KT 9999 SCT025 TEMPO 1515/1602 7000 -RA BKN014 PROB40 TEMPO 1516/1524 14018G32KT PROB30 TEMPO 1516/1521 4500 RASN FZRA BKN009 BECMG 1601/1603 22010KT PROB40 TEMPO 1603/1606 23015G25KT 7000 SHRA=

Typically, an aircraft may encounter rain ice (also known as ***'freezing rain'***) if it is flying in rain ahead of a warm front when the OAT is less than 0°C. Rain ice can also affect aircraft on the ground, and even form on cold runways and taxiways (when it may be referred to as 'black ice').

The only safe way to deal with rain ice and the like is avoidance – that is avoiding putting an aircraft in conditions that can lead to rain ice (freezing rain).

Frost

Tiny ice crystals that form on a stationary surface which has been cooled below 0°C are known as 'frost', often specifically **Hoar frost**. Hoar frost can form on a stationary aircraft either when dew freezes, or water vapour freezes directly onto the surface without going through the liquid state – a process called sublimation.

Figure M10.7
Hoar frost on an aircraft which had been parked outside in overnight sub-zero temperatures.

The ideal circumstances for the formation of hoar frost exist when an aircraft is parked in the open on a clear, cold night when there is radiative cooling (heat is radiating into space) and the air temperature is expected to fall below or close to freezing (0°C). In the morning the aeroplane surfaces may be covered in a delicate layer of frost. This is extremely dangerous if not removed. Although the frost weighs hardly anything, it will have a dramatic effect on the aerodynamics of the aircraft. Even a thin layer of frost will act like any other form of airframe icing – it reduces lift, increases drag, raises the stalling speed and can make the aircraft difficult to control or even unflyable. Performance is also reduced, leading to longer take-off and landing distances, reduced climb rates, slower cruising speeds and reduced range. It follows that, before flight, all frost must be removed from all the aircraft surfaces.

Figure M10.8
Airliner de-icing taking place at an airport in freezing conditions (note the green de-icing fluid on the wing). Commercial operators have access to sophisticated de-icing equipment and have comprehensive guidance on de-icing measures including fluids used, 'hold-over' times between de-icing and take-off etc. Most private pilots have no such support and the only safe advice is to ensure that all ice, frost or snow is removed from the aircraft before take-off.
Out of interest, on this occasion and despite two trips to the deicers, this commercial flight was cancelled due to the rain ice (FZRA) conditions.

It is also the case that any snow lying on the aircraft must be removed from all surfaces before flight. It is sometimes thought that snow will blow off an aircraft during the take-off, but this is not true. Any snow on the aircraft must be completely removed before flight. So, for example, if an aircraft is caught in a snow shower whilst awaiting take-off and the snow settles on it (especially over any wing fuel tanks, which tend to be cool) the pilot must remove all the snow from all the aircraft before take-off.

In addition to frost which occurs at the surface, in some circumstances Hoar frost can form on an aircraft in flight. This may occur, for instance, if an aircraft has been flying for a long time in sub-zero air (OAT below 0°C) and then descends into an area of warmer and possibly more moist air. Hoar frost can also form on an aircraft if it takes off when the surface air temperature is below freezing and then climbs through an inversion into warmer air. However, in this instance the frost normally thaws quickly.

In addition to the frost situations described above, conditions such as freezing drizzle and freezing fog are likely to lead to ice formation on an aircraft.

Piston Engine Icing

The conditions associated with airframe icing can also cause engine icing. Ice may form on the leading edges of propeller blades of an aircraft (or the rotor blades of a helicopter) and greatly reduce their efficiency. The resulting mechanical imbalance also results in considerable vibration. Furthermore, as already noted, flight in freezing precipitation such as snow can block or restrict the engine air intakes. Since flight in known icing conditions is not permitted unless the aircraft has de-icing equipment, avoiding such conditions should rule out the chances of encountering this type of engine icing.

However, by far the most common form of ice-related problem which afflicts a piston engine is **carburettor icing**.

When air enters a carburettor, the combined effect of the fall in pressure within the venturi and the evaporation of the fuel mixing with the air leads to a very large decrease in the temperature; a temperature drop of around 25 to 30°C is normal. In these conditions, water vapour almost inevitably forms into ice crystals which stick inside the venturi of the carburettor, constricting it and causing a reduction in engine power.

Because of the very marked reduction in temperature inside the venturi, it is not necessary for the outside air temperature (OAT) to be below 0°C for carburettor icing to occur. In fact, the dominant factor in carburettor icing is the moisture content of the air. The more humid the air, the greater the risk of carburettor icing. Warm air can hold more moisture than cold air, so it becomes evident that an engine is more likely to experience carburettor icing on a warm, humid and cloudy summer day than on a cold, dry, clear winter one. The obvious danger signs of carburettor icing conditions are visible moisture such as clouds, fog and precipitation; a moist airflow such as that from over the sea or a large body of water; or localised effects such as wet or water-logged airfields. Carburettor icing is a particular danger at low power settings.

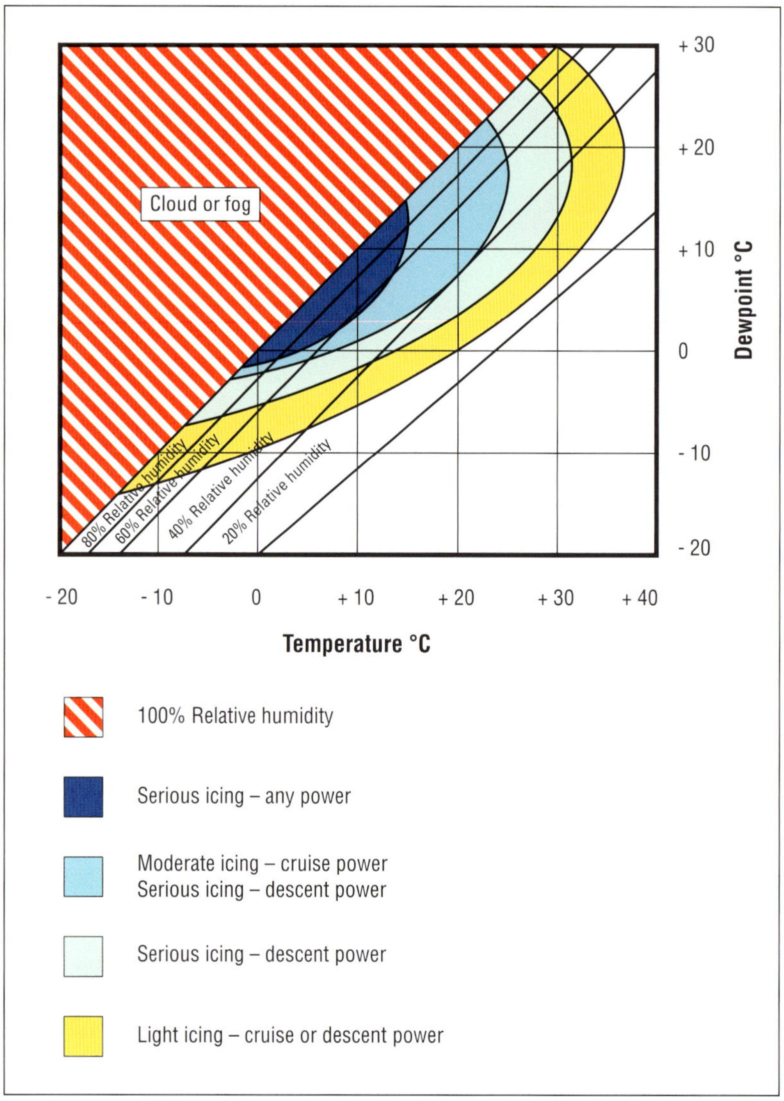

Figure M10.9
Air temperature and dew point (and therefore relative humidity) are key determinants in the likelihood of carburettor icing.

Graphs showing the conditions most likely to lead to carburettor icing tend to concentrate on the temperature and humidity. At OATs above +30°C, even the large temperature reduction in the carburettor should not take the air temperature in the venturi below freezing. At air temperatures below -10°C, any moisture is likely to become dry ice crystals which pass through the engine rather than sticking to anything. However, this leaves a wide range of temperatures when carburettor icing is possible, and indeed one of the biggest problem with carburettor icing is that it can occur over such a wide range of atmospheric conditions.

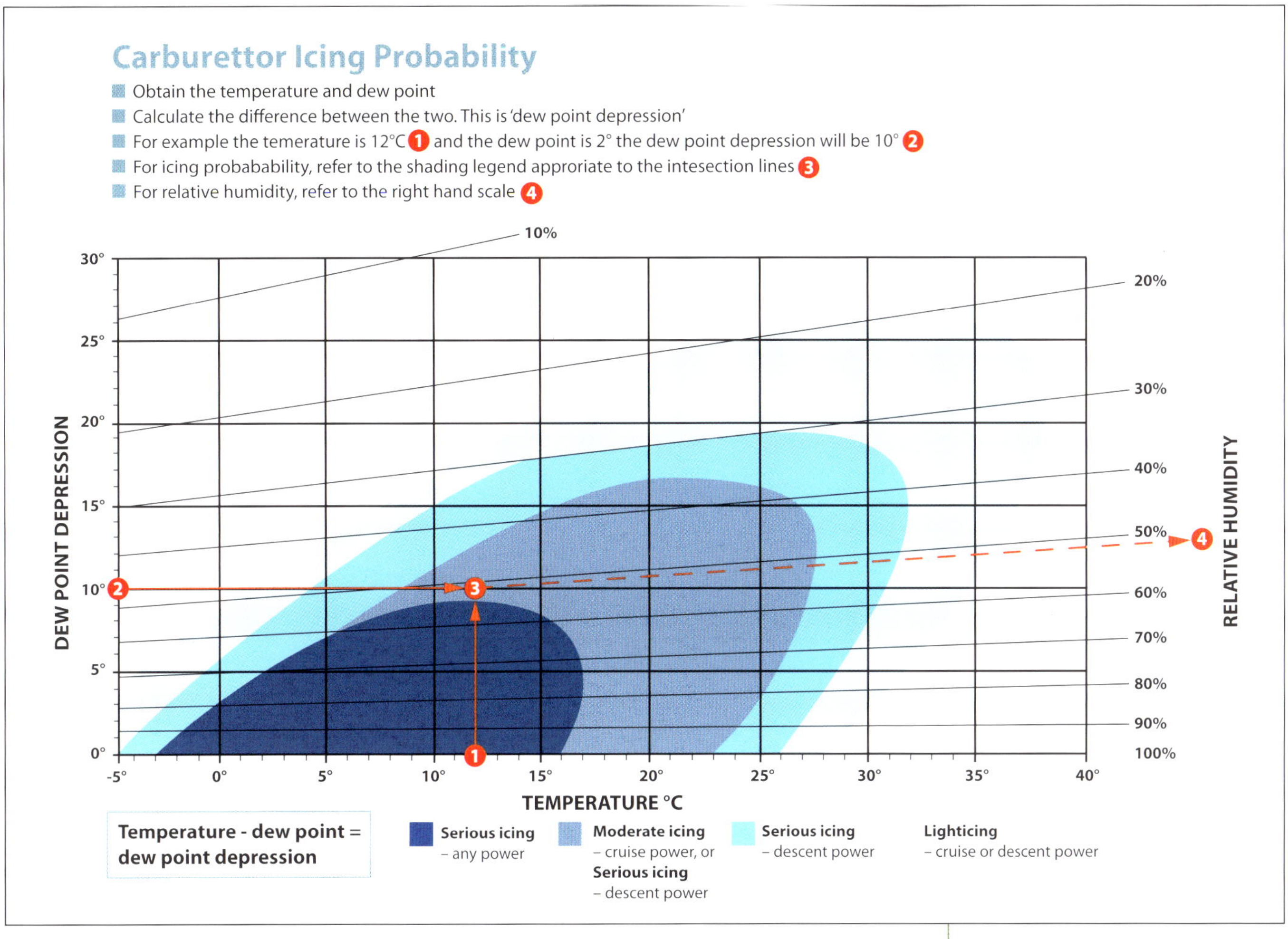

Even though the causes if carburettor icing are well understood and widely publicised, carburettor icing is still the chief suspect in a large number of piston-engine failures and it follows that all pilots of light aircraft with carburetted piston engines should be particularly knowledgeable about this subject. Information and advice about carburettor icing will be usually be found in the aircraft's flight manual and in safety information published by safety bodies and aviation authorities.

Figure M10.10
Many aviation authorities and safety organisations produce guidance material on the subject of piston engine icing.

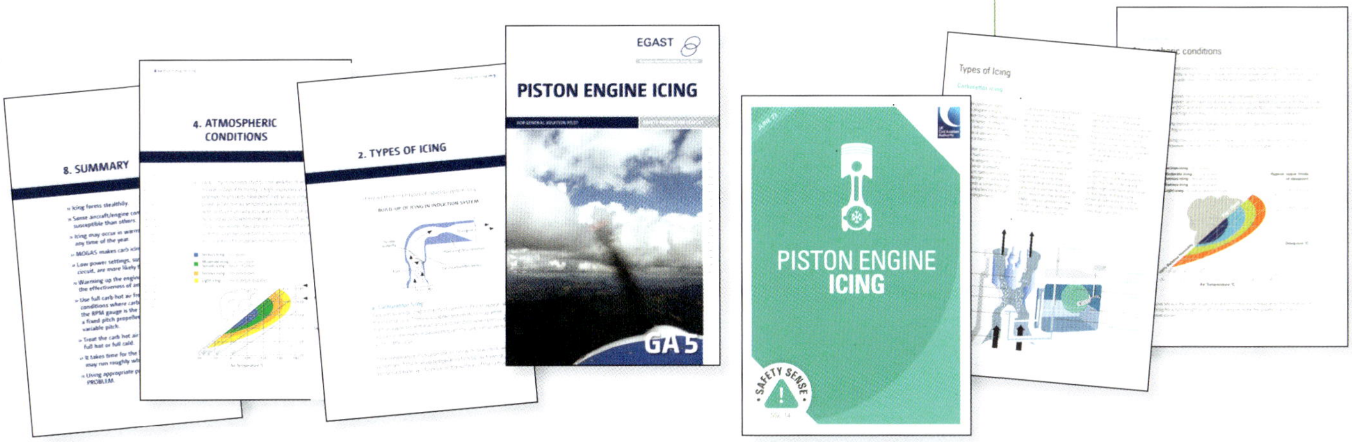

Progress Check

55. In general, what range of air temperatures are associated with the highest risk of airframe icing?
56. What name is given to a brittle and opaque layer of ice which forms on the leading edges of the wing and tail surfaces, on struts and radio aerials?
57. How can a pilot recognise 'clear' ice?
58. List at least five hazards associated with flight into icing conditions in a non-de-iced aircraft.
59. In which conditions is an aircraft most likely to encounter rain ice?
60. What is the meteorological code for rain ice?
61. Is a pilot required to take any action if the aircraft is found to have a layer of frost during the pre-flight inspection?
62. Is it necessary for the outside air temperature to be below 0°C for carburettor icing to occur?
63. At what engine power setting is carburettor icing most likely to occur?

These questions are intended to test knowledge and reinforce some of the key learning points from this section. In answering these questions, a 'pass rate' of around 80% should be the target.

Model answers are found at page M175-M176.

M11

Hazardous Weather Conditions: Thunderstorms

Formation of Thunderstorms

Hazards for Aircraft

Progress Check

M11 Hazardous Weather Conditions: Thunderstorms

Formation of Thunderstorms

Figure M11.1
A thunderstorm contains almost all the most hazardous weather conditions a pilot can encounter.

A fully-developed **thunderstorm** is one of nature's most impressive performances – indeed a thunderstorm contains nearly all the weather hazards an aviator should be most concerned about. Within this maelstrom, containing (it is said) the energy equivalent of several nuclear bombs, a pilot can find:

- **Winds** – strong enough to overturn a parked aircraft.
- **Rain** – heavy enough to flood an area in minutes.
- **Hail** – capable of denting metal and shattering windscreens.
- **Lightning** – vastly more powerful than mains electricity.
- **Turbulence** – severe enough to destroy even a large aircraft in mid-air
- **Wind shear and downdraughts** – capable of downing an airliner.
- **Icing** – severe anywhere above the freezing layer (contrary to normal icing theory).
- And even **tornadoes and waterspouts**.

Not surprisingly, the universal advice concerning thunderstorms is simple – keep well away! And this golden rule of avoidance is relevant to the pilot of *any* aircraft, whether it is a modern and fully-equipped airliner or a single-seat homebuilt.

There are three essential factors for thunderstorm development:

- An *unstable atmosphere*, with the instability reaching to the upper layers.
- A supply of *moist air*.
- Some form of *lifting action* to trigger-off the thunderstorm.

As a reminder, the stability of the atmosphere describes the tendency of the air to resist vertical motion**. Stable air** resists vertical motion. If air in a stable atmosphere is displaced vertically, it will tend to return to its original level. **Unstable air** does not resist vertical motion. If unstable air is displaced vertically, it may well continue to move – forming strong vertical air currents.

Figure M11.2
A 'stable' atmosphere can be likened to a calm sea, with little vertical movement and smooth conditions.

Figure M11.3
An 'unstable' atmosphere can be likened to a rough or stormy sea, with a lot of vertical movement and rough conditions.

In this sense the atmosphere can be likened to the surface of the sea. A calm sea has little vertical movement, just like stable air. A rough sea can have large vertical movements, just like unstable air.

To recap, temperature lapse rate in the atmosphere is the rate at which temperature changes with a change in height. An average figure of 2°C per 1000ft has been used so far. However, it is necessary to appreciate that the actual lapse rate is variable and changes all the time according to atmospheric conditions.

A knowledge of the actual temperatures throughout the atmosphere is important to weather forecasters. To collect this information, traditionally 'radiosonde' balloons are launched which transmit various items of information, including temperature, as they climb through the atmosphere. Increasingly, upper air observations from aircraft and satellites are also available to forecasters. Armed with this data, a forecaster can see the actual rate of temperature change with height – commonly referred to as the **Environmental Lapse Rate** (**ELR**). The ELR is then compared with average lapse rates for air that is moving vertically.

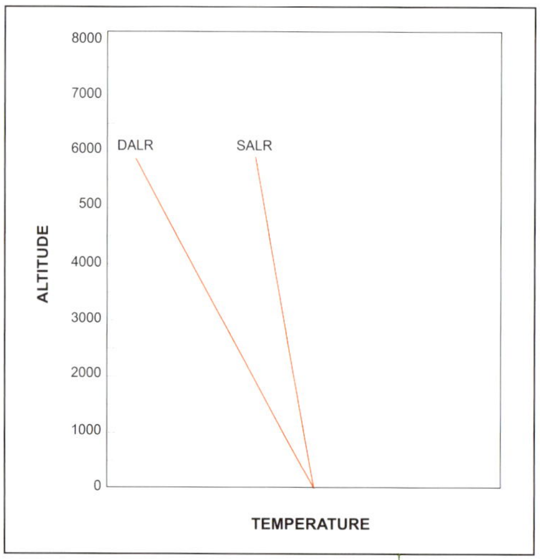

Figure M11.4

When rising, dry air is considered to cool at 3°C per 1000ft – the Dry Adiabatic Lapse rate (DALR). When saturated air is rising it is considered to cool at 1.5°C per 1000ft – the Saturated Adiabatic Lapse Rate (SALR).

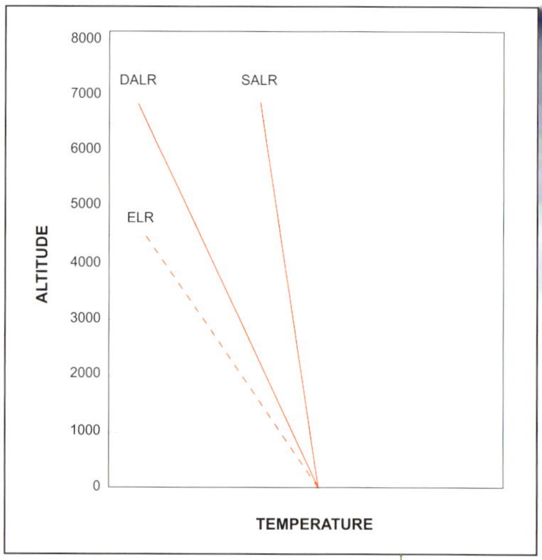

Figure M11.5

If the Environmental Lapse Rate (ELR) is greater than both DALR and SALR, the atmosphere can be said to be absolutely unstable. This atmosphere does not resist vertical movement.

Figure M11.6

The extensive vertical development of these clouds shows that the atmosphere is unstable. This is generally true when clouds grow taller than they are wide.

When dry air rises, its lapse rate is assumed to be around 3°C per 1000ft. This value is sometimes called the **Dry Adiabatic Lapse Rate** (**DALR**).

When saturated air rises its lapse rate is less than DALR, because as water vapour condenses into liquid droplets, heat energy is released. This reduces the rate of cooling, and so rising saturated air cools more slowly than dry air – at a rate of around 1·5°C per 1000ft. This is sometimes called the **Saturated Adiabatic Lapse Rate** (**SALR**).

By comparing the actual temperature profile of the atmosphere (the ELR) with the known DALR and SALR of rising air, a forecaster can make assumptions about the **stability** of the atmosphere.

If the actual temperature lapse rate in the atmosphere, the ELR, is less than SALR (that is, less than 1.5°C per 1000ft) the atmosphere can be described as 'absolutely stable'.

If the actual temperature lapse rate (ELR) in the atmosphere is greater than DALR (that is, greater than 3°C per 1000ft) the atmosphere can be described as 'absolutely unstable'.

If the temperature lapse rate in the atmosphere is between SALR and DALR, the atmosphere can be described as 'conditionally unstable'.

Temperature and humidity have a marked effect on atmospheric stability. Imagine a sunny summer day over land. Over a spot warmed by the sun, a parcel of quite dry air is heated above the temperature of the surrounding air. Because it is warmer than the surrounding air, it is less dense than the surrounding air, this makes it 'buoyant' and so starts to rise (like a bubble of less dense gas rising through a more dense liquid in a fizzy drink). As the bubble of air rises, it cools. If the air is dry, its **temperature lapse rate** will be around 3°C per 1000ft (this is known as the **Dry Adiabatic Lapse Rate**). Meanwhile, assume that the surrounding air is cooling at a different lapse rate (the **Environmental Lapse Rate**) of about 2°C per 1000ft. As long as the rising air is warmer (and less dense) than the surrounding air, it will keep rising:

In this example, at 5000ft the parcel of air reaches the same temperature as the surrounding air. It therefore loses its buoyancy and stops rising.

This example refers to a shallow layer of instability caused by convection. However, the atmosphere can be unstable (or stable) through a much greater depth. Stability is generally affected by heating or cooling of the atmosphere by the surface. Heating from below tends to make the atmosphere unstable. An unstable atmosphere is characterised by 'choppy' flying conditions, and convective clouds with significant vertical development

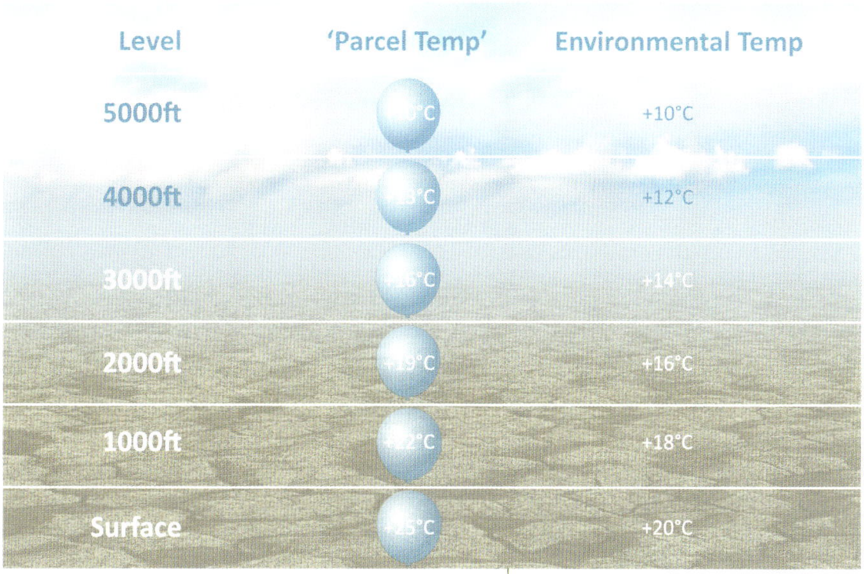

Figure M11.7

Imagine now that the rising air is very moist, rather than being dry, so that its dew-point temperature is about +20°C. Condensation will occur at around 2000ft. Once condensation is occurring, the parcel of air is saturated and so it cools at a reduced rate (say 1·5°C per 1000ft – the **Saturated Adiabatic Lapse Rate, SALR),** whilst we'll assume that the Environmental Lapse Rate – ELR – remains constant at about 2°C per 1000ft. Therefore, the rising parcel of air remains warmer (and less dense) than the surrounding air as it climbs higher into the atmosphere:

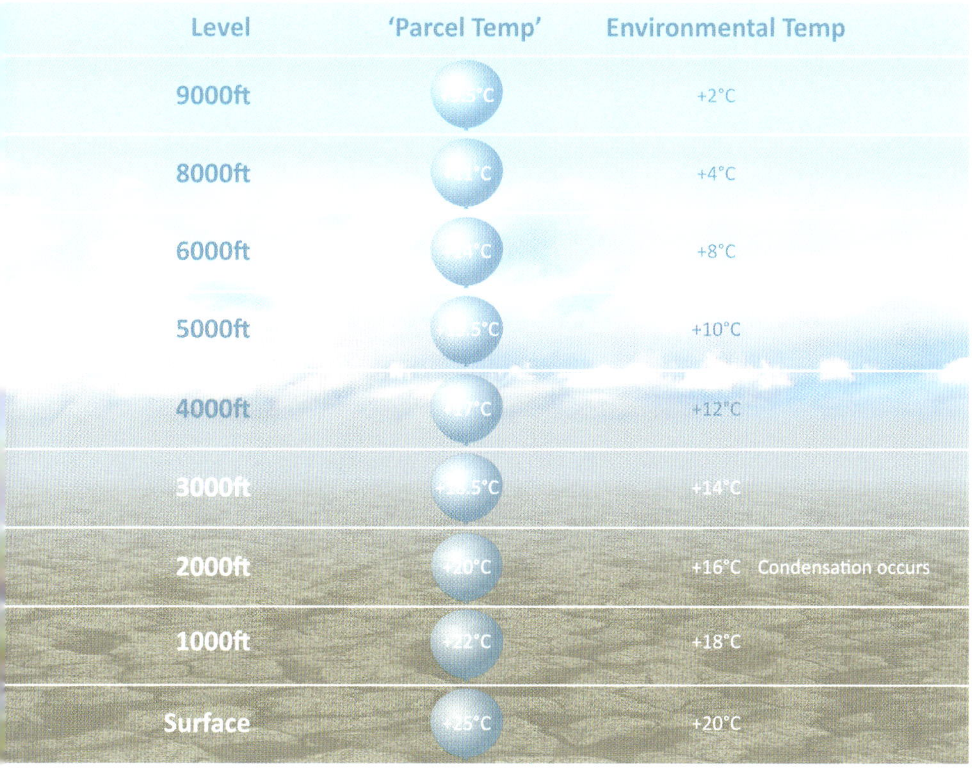

Figure M11.8
The rising parcel of dry air cools at around 3°C per 1000ft to around 2000ft, where it reaches its dewpoint and condenses into cloud. After that point, it cools at just 1.5°C per 1000ft, and it remains warmer than the surrounding air as it climbs.

The fact that the parcel of air is more humid than the surrounding air also makes it less dense and liable to remain more buoyant. This is an unstable situation, and in this example the condensing air releases more energy into the atmosphere as it climbs. Hence, the rising parcel of air is probably going to accelerate because because of its increased buoyancy over the surrounding air.

A plentiful supply of moisture in the air will supply the energy to stoke up a thunderstorm, by virtue of the process of **condensation**. Warm air can hold more moisture than cold air – so the warmer the air, the more powerful any thunderstorm is likely to be. Indeed, the capacity of air to hold water vapour increases sharply with increased temperature; air at +30°C can hold nearly three times as much water vapour as air at +15°C. On a stiflingly hot day when even the slightest task seems to make you break out in a sweat, the humidity is probably high. If other conditions are favourable, serious thunderstorms may be imminent.

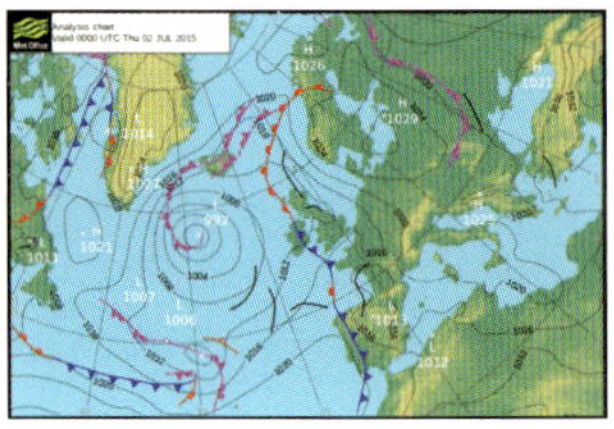

Figure M11.9

A European synoptic chart on a day when thunderstorms were forecast (and did in fact occur). It is mid-summer, a cold front is approaching, there is a depression over Spain, multiple troughs, and warm unstable air is being fed up from the south.

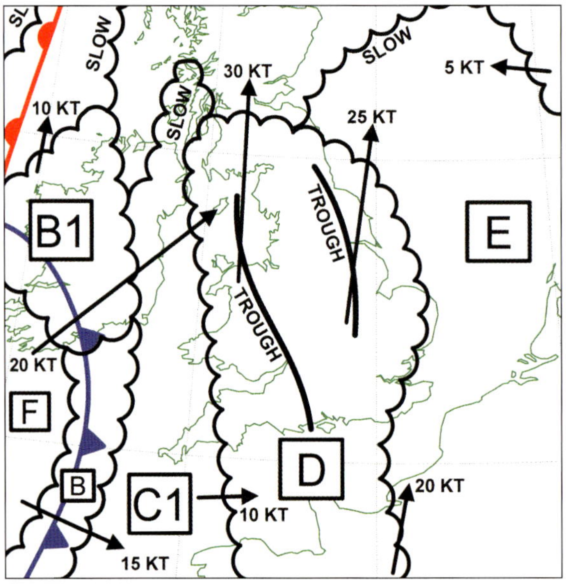

Figure M11.9a

Another weather chart for the same day. Thunderstorms were forecast (and did occur) around the troughs in area D.

Figure M11.10

A relatively shallow layer of instability in the middle layer of the troposphere is marked by these Alto Cumulus (AC) clouds. Alto Cumulus clouds like these can give several hours warning that thunderstorms are likely.

Figure M11.11

Another view of a sky full of Alto Cumulus; extensive thunderstorms occurred later in the day.

Figure M11.12

A 'squall line' – an organised line of Cumulonimbus (CB) clouds.

Given both instability and high humidity, the final ingredient n thunderstorm formation is a lifting action. Lifting at a front can set off thunderstorms. The obvious candidate is a cold front which, classically, tends to form cumulus-type clouds. However, extensive cumulonimbus (the cloud type associated with thunderstorms) can form at warm fronts in some conditions, although more rarely. If anything these are more dangerous, because they tend to be embedded in other clouds associated with the warm front. Likewise, although the lifting action at a cold front might appear to be an obvious trigger of thunderstorms, in summer thunderstorms are more often set off well ahead of a cold front – sometimes in an organised line called a **squall line**. This line of thunderstorms brings all the hazards expected of thunderstorms, but when the actual cold front arrives (perhaps many hours later) it is often weak and inactive. In fact, the vast majority of fronts that cross Europe contain no thunderstorms.

Aside from fronts, other forms of atmospheric lifting action can set off a thunderstorm. **Convection** is an obvious one, causing thunderstorms to form over surface 'hot spots', especially those which have been heated by the sun. Most summer thunderstorms are triggered in this way, usually in anticyclonic situations when intense surface heating eventually breaks down the naturally stable nature of an anticyclone. An unstable upper airflow will also encourage thunderstorms to form, but this is for the forecasters to spot. **Orographic lifting**, in which the air is forced to rise by terrain, is also an effective trigger action for thunderstorms, and the combination of mountainous terrain and a thunderstorm is perhaps one of the most lethal flying situations a pilot can encounter.

Thunderstorms can also be set off by **convergence**, for example where a sea breeze moving inland (and carrying warm moist air) meets a general airflow coming the other way.

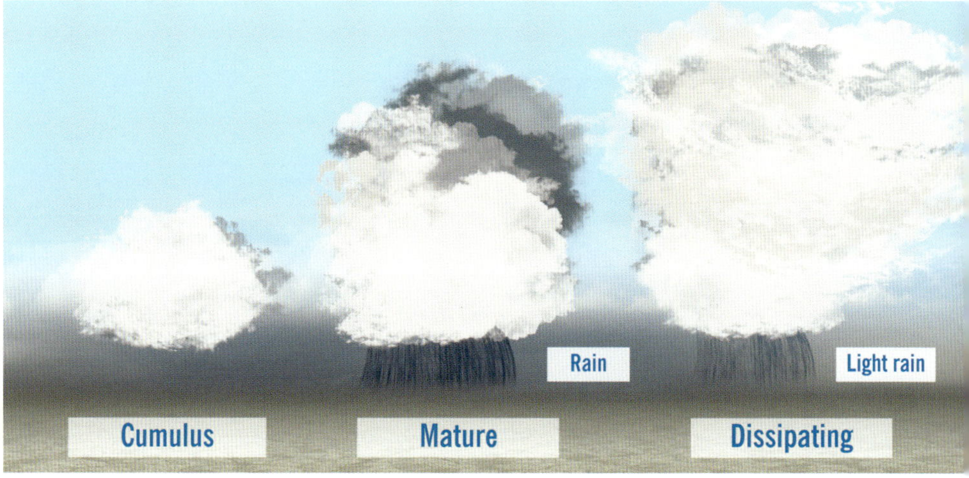

Figure M11.13
The three stages of a thunderstorm's 'life-cycle'.

Once lifting has started, the first visible sign of a developing thunderstorm is the formation of cumulus cloud. This is the so-called **cumulus stage**. What distinguishes a baby thunderstorm-type cumulonimbus from an ordinary cumulus is vertical growth. The top of the cumulus cloud keeps on climbing at an astonishing rate as massive condensation releases energy to keep the cloud rising and the process going. Once a cumulus is taller than it is wide it can be described as a **towering cumulus**, abbreviated in weather reports and forecasts to **TCU**. Not all towering cumulus develop into a cumulonimbus, but many do. This stage of development commonly sees vertical currents of 3000ft per minute, although currents of up to 6000ft per minute are not unknown. This gives the first rule thunderstorm avoidance: *don't* try to out-climb one. A growing cumulus will simply out-climb most aircraft and may well envelope it. At this stage the cumulus has distinct 'hard' edges and a bright white brilliance; it is not unlike a cauliflower in appearance. Not all cumulus clouds that look like this will become thunderstorms, but it is sensible to avoid these 'hard-edged cumulus', even if thunderstorms are not forecast.

The **mature stage** of a thunderstorm is marked by precipitation beginning to fall from the cloud. The precipitation is likely to be heavy, and it is at this stage that the vertical currents in and around the cumulonimbus are at their most intense. The rain (or snow or hail) can be accompanied by hail and lightning. The cold downdraught from the storm may form a **gust front** pushing out ahead of the storm.

The **dissipating stage** marks the beginning of the end of the storm. As the top of the cloud reaches a marked inversion (probably the tropopause), the ice crystals forming the head of the cloud can spread out laterally, blown by the upper winds into the characteristic 'anvil' shape. The word 'dissipating' is a relative term, the thunderstorm can still be dangerous for some time after the anvil has formed. Although the active stage of a thunderstorm typically lasts less than an hour, the cloud may persist for two or three hours or more.

Figure M11.14

Towering Cumulus (TCU) – cumulus clouds that are taller than they are wide.

Figure M11.15

Towering Cumulus beginning to grow larger to become Cumulonimbus (CB).

Figure M11.16

A growing CB, it has no anvil top yet but is dangerous nevertheless.

Figure M11.17

A thunderstorm at the 'mature' stage.

Figure M11.18

A gust front pushing ahead of a thunderstorm, caused by the outflow of cold air from the thunderstorm.

Figure M11.19

A line of 'roll cloud' marking the gust front pushing ahead of a thunderstorm – in effect a mini cold front.

Figure M11.20

This cumulonimbus has reached the tropopause and begun to spread-out leading to a classic 'anvil' top.

EGGD BRISTOL AIRPORT

TAF EGGD 231657Z 2318/2418
12008KT 9999 FEW040
TEMPO 2318/2322 4000 +TSRA
TEMPO 2322/2402 3000 +SHRA TSRA
PROB30 TEMPO 2322/2402 VRB15G25KT +TSGR
BECMG 2400/2403 27012KT
TEMPO 2402/2407 3000 BR BKN003
PROB30 TEMPO 2403/2406 0300 -DZ FG
BKN001
PROB30 TEMPO 2407/2409 7000 BKN008
BECMG 2412/2415 21006KT=

Figure M11.21
A truly horrible TAF aerodrome forecast! Of particular note are the lines:
TEMPO 2318/2322 4000 +TSRA (temporarily between 18:00UTC on the 23rd and 22:00, visibility 4000 metres in heavy thunderstorms)
TEMPO 2322/2402 3000 +SHRA TSRA (temporarily between 22:00UTC on the 23rd and - 02:00 on the 24th, visibility 3000 metres in heavy showers of rain and/or thunderstorms with rain)
PROB30 TEMPO 2322/2402 VRB15G25KT +TSGR (30% probability that temporarily between 22:00UTC on the 23rd and 02:00 on the 24th, surface wind variable at 15 knots gusting 25 knots, heavy thunderstorms with hail).

This brief description applies to the typical 'single cell' storm. When precipitation begins, it falls through the updraughts within the cloud. The cooling effect damps down the vertical motion and will eventually stop it, thus bringing the storm to an end. However, in certain circumstances – particularly when the winds aloft are much stronger than the winds lower down, the whole storm may 'tilt'. This means that the falling precipitation misses the updraughts and the storm can become self-propagating. Sometimes described as 'severe local storms' or 'supercells', these storms will be more severe and longer lasting than an 'ordinary' thunderstorm. Although often associated with North America, such thunderstorms are not unknown in western Europe. Furthermore, the life-cycle described relates to a single thunderstorm 'cell'. It is quite possible for a cumulonimbus cloud to contain more than one cell, each at a different stage of development. This will prolong the duration of the storm.

Statistically, thunderstorms in Europe are most frequent in the summer months when convection or orographic uplift triggers them. Moving south towards the tropics, thunderstorms become more frequent and more severe. This is chiefly due to the increased temperature and humidity of the air; the higher tropopause in these latitudes also allows the storm greater vertical development.

Hazards for Aircraft

As stated at the beginning of this chapter, thunderstorms can contain – in a single package – many of the most destructive weather conditions known to aircrew. From pilot's point of view, a short summary of these hazards is useful:

Rain and snow
Any precipitation from a thunderstorm is likely to be heavy. An intense downpour of rain may reduce visibility to a few kilometres or less, and the high humidity increases the risk of carburettor icing. Falling snow has an even more dramatic effect on visibility. In very heavy snow, visibility may be as low as 50m.

Figure M11.22
A thunderstorm arriving over an airfield – the last possible moment to secure the aircraft and take shelter.

Figure M11.23
Intense rain under a thunderstorm, leading to 'flash floods' at the surface and greatly reduced visibility.

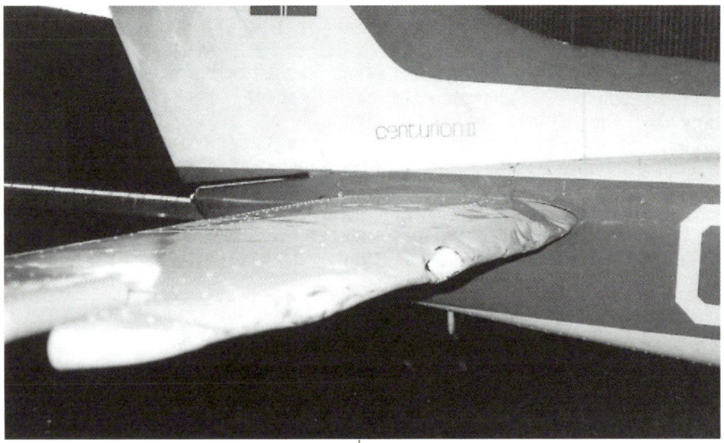

Figure M11.24a
Damage inflicted on a light aircraft after flying through hail.

Hail
Precipitation in the form of hail is worth separate consideration. Hail forms at first as small ice stones. However, if a falling hailstone meets a strong updraught, it can be carried back up into the storm and so acquire another layer of ice. This may repeat several times – if you cut through a hailstone you can count the layers and see how many times it has been 'recycled' in the storm. Hail can cause serious damage to aircraft by battering leading edges, shattering windscreens, destroying aerials and so on. The fall of hail is difficult to predict, and may occur outside the main cloud from the overhanging anvil. Hail can even be thrown out of the side of a thunderstorm – another good reason for staying well away from an active cumulonimbus cloud. Some pilots believe that thunderstorms containing hail have a peculiar greenish tint, although it is a subjective judgement.

Figure M11.25
An area on 'mammus' under an active thunderstorm, marking an area of turbulence and strong downdraughts.

Turbulence
There will inevitably be severe turbulence in and around an active thunderstorm. An aircraft may encounter an updraught of more than 4000ft per minute, followed immediately by a downdraught of a similar magnitude. Such turbulence is in itself strong enough to destroy an aircraft in mid-air, and several instances of this are known. Moreover, particularly in respect of small aircraft, even turbulence of a lesser magnitude can still be enough to cause loss of control; the aircraft may then be overstressed in the subsequent recovery manoeuvres. Sometimes 'mamma' clouds (egg-shaped protrusions in the base of the cloud) can be seen beneath a thunderstorm, which mark an area of downdraughts and turbulence. Avoid this area.

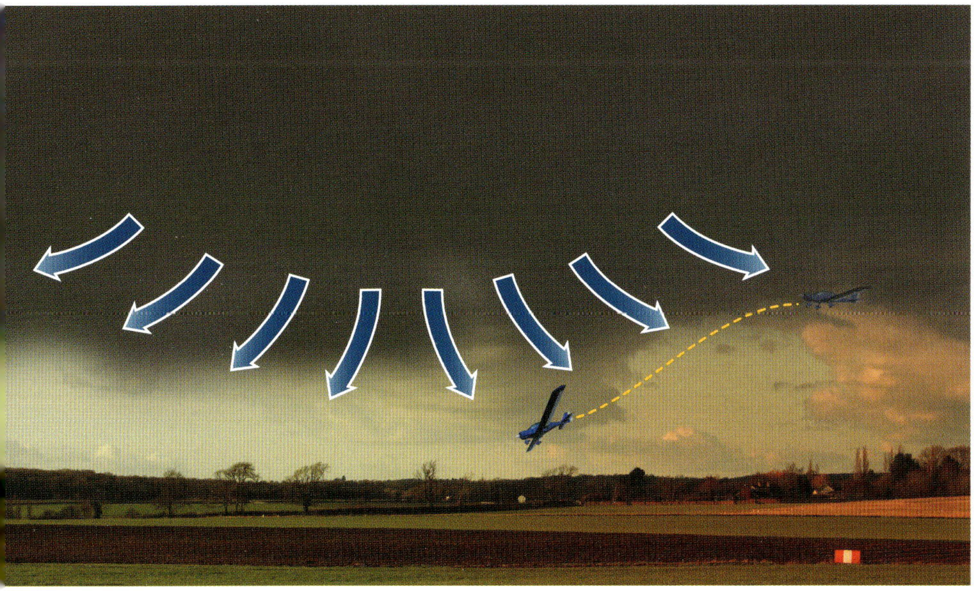

Figure M11.26
Strong downdraughts under a thunderstorm can lead to gale-force winds and severe wind shear.

Wind shear, Downdraughts and Microbursts

Thunderstorms can produce all these phenomena, which are particularly hazardous to aircraft taking-off and landing. Downdraughts often occur under the thunderstorm, where a column of rapidly descending air hits the ground and spreads out. An aircraft flying into such a downdraught may first encounter an increased headwind, followed by the unannounced onset of wind shear (a sudden charge of wind velocity over a short horizontal or vertical distance) and possibly a tailwind leading to a loss of airspeed. A *microburst* is a very localised and intense downdraught. Whilst downdraughts are usually associated with heavy rain, microbursts are often 'dry'. The general advice is never to take-off or land into the precipitation from a thunderstorm, it is best to wait until the storm has moved well away. **Virga** – rain or snow that evaporates before reaching the ground – may also indicate an area of strong downdraughts.

The gust front occurring up to twenty miles ahead of a thunderstorm (rather like a mini cold front) can cause severe wind shear, as the general airflow is replaced by the cold outflow from the storm. The gust front will probably be an area of turbulence which may be marked by **roll cloud**. The sudden change in wind direction, increase in wind speed, and coldness of the air associated with a gust front can be a final warning that the arrival of a thunderstorm is imminent.

Strong winds

Even though the general wind flow around a thunderstorm may be light, a thunderstorm itself can produce gale-force winds. The thunderstorm outflow (the cause of the gust front) may have a wind speed of up to 40kts. A downdraught or microburst may give gusts of 60kts or more as it hits the ground.

Figure M11.27
A roll cloud that appeared ahead of a particularly violent thunderstorm.

Figure M11.28

The aftermath of a severe thunderstorm (and probably associated mini-tornado) which passed through an airshow in Florida (the author had left the show a few minutes before the thunderstorm arrived).

Figure M11.29

Although smaller and less common than in the USA for example, small tornados do occur in Europe in association with active thunderstorms and a tornado has destroyed at least one airliner in European skies. A tornado may be visible as a 'funnel cloud' under a thunderstorm, and as a waterspout if a tornado touches down on water.

Tornadoes

A **tornado** is a column of rotating air, bred within a thunderstorm and giving wind speeds of up to 200kts. A tornado may be completely enclosed within the cloud and therefore invisible to the pilot. Alternatively, a tornado can be visible as a **funnel cloud** underneath the thunderstorm, or even appear as a **waterspout** where it touches down on water. Although often thought of as a North American phenomenon, tornadoes do occur in Europe – where they are quite capable of taking the roofs off houses, destroying caravans and snapping trees. Technically, a tornado only occurs when it touches a land surface, but this is semantics. Very few aircraft, whether on the ground or in the air, will survive an encounter with a tornado, water spout or funnel cloud.

Figure M11.30

Lightning is a lot easier to see and avoid at night.

Lightning

Metal aircraft usually survive a lightning strike (although a few have been destroyed, apparently by the ignition of fuel vapour), with damage to the aircraft sometimes limited to holed or blown-off wing-tips. After a lightning strike, the radios and other avionics may be unusable and the compass will certainly be unreliable. Aircraft constructed from wood, composites or GRP tend not to have such good electrical bonding as their metal counterparts, and they may not escape a lightning strike so lightly.

Lightning and the general electrical activity in and around a thunderstorm will lead to increased radio 'static' and interference, particularly at the lower frequencies such as those used by an Automatic Direction Finding (ADF) receiver. An ADF needle may well point directly at the nearest active thunderstorm cell, regardless of the location or frequency of the radio beacon to which it is tuned. The VHF portion of the radio spectrum, as used by communication and VOR/ILS radios, is usually less afflicted. It is generally considered that GNSS receivers are unaffected by thunderstorm activity, although caution in this respect is prudent.

On the ground, it is possible to establish approximately how far away a storm is by estimating the time interval between a lightning flash and the associated clap of thunder (counting *"Thousand and one, thousand and two"* etc. usually works well). Each five-second interval between flash and bang equates to approximately one mile.

Icing

Within a thunderstorm, icing may be severe at any altitude above the freezing level and few light aircraft are certified for flight into known icing conditions. Severe icing is most likely to be encountered within the cumulonimbus cloud at air temperatures of between 0°C and -20°C.

By now, it should be obvious that the immediate vicinity of a thunderstorm is emphatically not the place to be in any aircraft – so anything which can minimise the possibility of an encounter with a thunderstorm is worth reiterating. The advice that follows is based on practical experience.

- The golden rule of thunderstorms is avoidance. When thunderstorms are forecast, consider your need to fly very carefully; plan and conduct any flight with avoidance in mind.
- Forecasters will often include thunderstorms in a forecast even if there is only a slight chance of them happening. Obtaining the most up-to-date area forecasts, as well as current actual reports and forecasts for departure, en route and destination airfields is vital. A thunderstorm may go through its entire life-cycle in less than 45 minutes, so a weather observation two hours old is of little use.
- Thunderstorms are likely to be particularly severe and violent if the dew-point is around +20°C or more.
- In northern Europe, a returning polar maritime air mass gives the greatest risk of thunderstorms, although when thunderstorms occur in tropical continental air they tend to be more severe.
- Thunderstorms embedded in other cloud, or those occurring in hazy or misty conditions, are very difficult to see! Flying above any haze layer, and staying clear of cloud, will improve flight visibility. If you are flying in haze or mist and it suddenly becomes dark, you may have flown into the shadow of a cumulonimbus. Immediate evasive action is advisable.
- Thunderstorms triggered by convection tend to be reasonably well separated and hence easier to avoid. Frontal, trough or line-squall thunderstorms are often close together or actually merging; for practical purposes these are impossible to fly around. Frontal, trough and line-squall thunderstorms also tend to be more persistent than convective ones.
- Even the smallest thunderstorm cell will be about five miles wide. The biggest can be up to 30 miles across.
- In flight, the weather information broadcasts help keep pilots updated on the general weather situation. Pilot or ATSU reports of cumulonimbus, or airliners asking to fly headings to avoid "*build-ups*" are warning signs.
- Keep well away from a developing cumulonimbus cloud or a full-blown thunderstorm – at least 10 miles from the edge of the cloud. Avoiding it by 20 miles is even better. Do not fly under, over or close to a developing cumulonimbus (CB).
- The more frequent the thunder and lightning from a CB, the more severe the thunderstorm.
- Many thunderstorm hazards (lightning, hail, wind shear etc.) exist outside the associated cloud. Turbulence may be experienced up to 20 miles downwind of the storm.
- Do not attempt to fly under a thunderstorm, even if there is no precipitation and the other side can be seen. There may be severe turbulence; precipitation may develop; and updraughts may even 'suck' the aircraft up into the cloud.
- Do not take-off with a thunderstorm overhead or in the vicinity. If a thunderstorm is over or near the destination airfield, go somewhere else or hold clear until it has moved well away. Do not attempt to land while the thunderstorm is close.
- Consider reporting any thunderstorms seen. Such real-time reports can be invaluable to other pilots in the area.
- Thunderstorms are also a hazard to aircraft on the ground. The safest place for an aircraft is inside a hangar.
- Finally, much of this advice applies to flying near large and growing cumulus clouds which may, or may not, become full-blown thunderstorms.

Progress Check

64. List the three main factors involved in the formation of thunderstorms.
65. What cloud type is described by the meteorological code TCU?
66. What cloud type is associated with an active thunderstorm?
67. List the three stages of an active thunderstorm.
68. What is the meteorological code for an active thunderstorm with heavy rain?
69. List at least five major hazards to aircraft associated with a thunderstorm?
70. What is the 'golden rule' for mitigating the risks associated with a thunderstorm?
71. Is it safe to fly under a thunderstorm if the pilot can see that Visual Meteorological Conditions exist to the other side of the storm?
72. Is it safe to take-off if there is an active thunderstorm overhead the departure airfield?

These questions are intended to test knowledge and reinforce some of the key learning points from this section. In answering these questions, a 'pass rate' of around 80% should be the target.

Model answers are found at page M176-M177.

M12 Other Hazardous Weather Conditions

Mountainous Areas

Turbulence

Wind Shear

Strong Winds

Progress Check

M12 Other Hazardous Weather Conditions

Mountainous Areas

Meteorology is often concerned with the large-scale state of the atmosphere, and influences that gradually alter weather patterns over areas of thousands of square miles. However, weather can also be modified by very localised or even isolated factors. Terrain is one such feature, and mountainous areas are particularly good at altering the weather and causing localised conditions.

Figure M12.1
Pilot's sometimes refer to 'Cumulus Granitus' – the cloud with the rock centre. 'Scud running' around high terrain is a depressingly common way of finding such a cloud with the ensuing high risk of Controlled Flight Into Terrain (CFIT).

Rising ground forces air to rise, even if it had no inclination to do so otherwise. If air is lifted (and hence cooled) far enough, it will reach its dew-point and cloud will form. The situation in which cloud covers high ground is far more dangerous than many inexperienced pilots realise. It is responsible for a great many weather-related accidents, where pilots fly aircraft into the ground simply because they never saw it coming.

Figure M12.2
Cumulus cloud forming over a mountain top due to orographic lifting.

Figure M12.3
Cloud partially covering high ground – best viewed from a safe altitude.

Figure M12.4
Cloud covering high ground, with poor visibility likely close to the cloud base.

The process of **orographic lifting** means that the cloud base around mountains and high ground can actually be lower than elsewhere. Likewise, the cloud base will fall quickly in any precipitation triggered by orographic uplift. Now imagine a VFR pilot trying to stay in visual conditions by flying just below the cloud base; a dangerous technique usually referred to as 'scud running'. With little forward visibility and possibly lost, the pilot presses on irrespective of the fact that the aircraft is being forced to descend – below the safety altitude – possibly straying in and out of the cloud base as the pilot dips lower and lower in an attempt to stay in visual contact with the ground. Meantime the ground may well be rising to meet the aircraft. This is the classic set-up for what accident reports refer to as 'Controlled Flight Into Terrain' (abbreviated to CFIT).

Figure M12.5
Airflow being 'funnelled' through a valley can have a direction and speed significantly different to the general wind.

As well as inducing cloud formation, mountains and valleys can also influence a general airflow by deflecting its direction, or by funnelling the wind through a narrow valley. Wind through a valley can be veered or backed up to 30° from the general wind direction, and have considerably higher speeds than the general air flow. Possibly the best-known **valley wind** in Europe is the Mistral, which can easily exceed gale-force as air is funnelled through the Rhone valley in southern France. The Tramotane, which affects the French and Spanish Mediterranean coast, is another example.

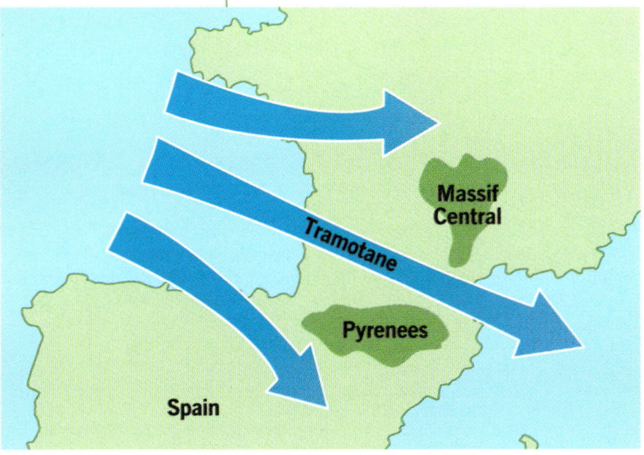

Figure M12.6
The often gale-force and turbulent Tramontane experienced along the Spanish and French Mediterranean coast is a local wind 'funnelled' between the mountains of the Pyrenees and the high ground of the Massif Central.

Figure M12.7
On a clear night, heat from the surface radiates into space, allowing the ground (and the air next to it) to cool. The cool air at the surface is more dense than the air above it, and so flows down the slope – a 'Katabatic' wind.

Figure M12.8
Early morning fog filling valleys whilst the peaks of the Alps remain clear.

On a more local scale, at night air cooling by contact with sloping ground will become more dense than surrounding air and flow down the slope – a **katabatic wind.** The reverse effect by day, when air is warmed on a sun-facing slope and, becoming, less dense flows up the slope, is known as an **anabatic wind**. In conditions favourable to radiation fog, the mountain tops often remain clear but fog can persist in deep valleys (which may stay in shadow) long after it has cleared elsewhere.

The wind blowing over a mountain range can also create large-scale disturbances in the atmosphere. These disturbances are usually referred to as **mountain waves** but are also known as 'standing waves' or 'lee waves' to glider pilots who make extensive use of them. Mountains can cause the air flow to undulate, like the swell in the ocean. Where mountain waves rise up, gliders have reached great heights with their assistance. However, where the waves turn down (sometimes called a 'down draft') the descending air can force aircraft to descend, even at full power and best climb airspeed. This situation can continue right down to ground level.

Figure M12.9
The downdraughts in the lee (downwind side) of high ground can be strong enough to defeat the best rate-of-climb of a light aircraft.

The basic conditions favourable to the formation of an organised mountain-wave pattern are:

- A wind of 15kts or more at the surface, blowing at right angles to (or within 30° of) the ridge or mountain-range axis, preferably with the wind strength increasing with height, but direction remaining fairly constant.
- A layer of stable air just above the ridge, bounded by less stable air. This condition may be indicated by an inversion just above the ridge (which is more likely in an anticyclone).

Figure M12.10
Lenticular clouds forming on the crest of mountain waves.

Figure M12.11a
Evidence of wave activity in high-level cirrus cloud.

Figure M12.11b
Wave clouds over the Scottish highlands.

Figure M12.11c
Small scale wave activity evident in the top of a cloud layer.

A mountain wave may not reveal its presence until the aircraft's rate of climb or descent begins to conflict with the power setting and attitude. When cloud does form in a mountain wave, it will often be of the **lenticular** variety. This is smooth and elongated cloud – not unlike an almond in appearance – which actually remains stationary at the 'crest' of the wave, the cloud forming at the leading edge and dissipating at the trailing edge. In favourable conditions there may be lenticular clouds at the crest of each wave, and maybe even several such clouds piled on top of one another.

Figure M12.12
The turbulent rotor which may be found just downwind of the crest of a hill may, or may not, be marked by roll cloud.

Flight within mountain wave itself can be exceptionally smooth. However, severe turbulence will be found in the **rotor** that can form downwind of the high ground beneath the crest of the waves. The rotor may be given away by a 'roll' (or 'rotor') cloud, a ragged cumulus that forms in the rotor zone. At first glance a roll cloud may look quite harmless. However, closer examination will reveal that it is rolling around its horizontal axis, revealing the violent turbulence in the rotor. Occasionally the roll cloud may form as a long tube just downwind of the ridge. The most extreme turbulence is usually found in the first rotor downwind of the ridge, and the ferocity of such a rotor is notorious. It is considered strong enough to destroy a powered aircraft and it has certainly destroyed gliders and caused aircraft to lose control. Even in light wind conditions when mountain waves and rotors are not forecast, the area just downwind of and below the level of a ridge or obstruction is a dangerous place to be. It is best avoided as a matter of course.

Obviously the first safety feature if planning to fly over or near high ground in conditions favourable for mountain waves is *height.* The higher above the mountains, the better. The basic rule of thumb is to double the elevation of the high ground and make that the minimum crossing altitude. Further, although crossing from downwind to upwind holds obvious risks, flight across high ground at an angle, or parallel to and downwind of a ridge, will leave an aircraft in wave conditions for much longer. Just as mountains and low cloud are dangerous, so mountains and strong winds also make a hazardous combination.

If the air flow over high ground is moist enough to give precipitation on the windward side of the mountain range, the downwind side will probably have a higher cloud base, higher temperatures and less precipitation. These are the result of the **Föhn effect.**

Figure M12.13
Föhn effect can cause drier and warmer conditions on the lee (downwind side) of high ground.

Föhn effect is a practical illustration of the difference between saturated and dry adiabatic lapse rates (SALR and DALR). If a moist airflow meets a mountain range and is forced to rise, the air initially cools at the DALR – say 3°C per 1000ft – until the air reaches its dew-point and cloud forms. If it continues to rise, the saturated air will now cool at the SALR, about 1·5°C per 1000ft. As the lifting continues, precipitation may fall on the windward side of the mountains. As the air passes the mountain peak, it begins to descend and therefore to warm. Having lost much of its moisture on the windward side of the mountain, the air soon warms past its dew-point and becomes unsaturated. Thereafter it descends and warms at the DALR of 3°C per 1000ft. The net result is that on the leeward (downwind) side of the mountains, conditions tend to be both warmer and dryer than on the windward (upwind) side. Areas on the leeward side of high ground are often said to be in the 'rain shadow' of the mountains. Just such an effect is observed in eastern Scotland, which in the prevailing westerly winds is on the leeward side of the mountains of the Scottish Highlands.

So high winds and mountain waves permitting, flight on the downwind side of the mountain range can often give better flying conditions (in terms of cloud base and visibility) than flight on the upwind side

Possibly the most severe mountain weather will be caused by a thunderstorm over high ground. Thunderstorms triggered by orographic uplift seem particularly violent, and when the assorted hazards of a thunderstorm are complemented by the dangers of high winds over mountainous terrain, reduced terrain clearance and so on, there is a particularly dangerous combination of flying conditions. It is also worth knowing that cloud formed by orographic lifting will almost certainly cause more severe icing than non-orographic cloud

Turbulence

Turbulence is any disturbance of the air flow causing eddies and variations. For practical purposes, turbulence only becomes a problem when it affects the handling of the aircraft and/or causes discomfort to the aircraft's occupants. However, vertical gusts may also increase the load factor on the airframe (risking structural damage) and may even cause the wing to exceed its critical angle of attack, causing the wing to stall at a much faster airspeed than expected. Apart from extreme examples, such as that encountered in and around thunderstorms and rotors, there are generally two principal causes of low-level turbulence. It may be convective (thermal), caused by vertical air currents; or frictional (mechanical) caused by the flow of air over the ground.

Convective turbulence is at its worst in light winds when surface heating is strongest, although it can be very difficult spotting which areas will generate the strongest thermals and the most turbulence – ask any novice glider pilot. If the air is moist enough, rising convection currents (thermals) may be topped by cumulus clouds. Avoiding these will keep an aircraft clear of some convective turbulence, and also help to avoid gliders that tend to congregate in thermals.

Figure M12.14
Low level 'mechanical' turbulence may occur downwind of obstructions such as buildings and trees.

Mechanical turbulence is most marked in strong wind conditions over irregular terrain, particularly downwind of hills or obstructions. In either case, increasing altitude is usually a good way to reach smoother air.

The aircraft flight manual will state a recommended turbulence or 'rough air' speed to be used if turbulence warrants it. In turbulence the airspeed may fluctuate, and attempts to 'chase' the speed may make matters worse. It is usually better to maintain the correct attitude and power setting for the desired performance and accept minor and temporary fluctuations in airspeed.

Turbulence can be a problem during the final approach to land or just after take-off. In both cases the aircraft can be subject to the low-level effects of airflow around buildings, trees or local terrain. A slight (5-10kts) increase in approach speed may a worthwhile precaution if approaching to land in turbulent conditions, assuming landing distance required in relation to runway length is not critical. Pilots will often expect turbulence and wind shear at their local airfield when the wind blows from a certain direction, and it is a strange fact of an aviator's life that a strong crosswind often seems to produce the most pronounced turbulence and wind shear.

Wind Shear

Wind shear is a sudden change of wind velocity over a horizontal or vertical distance (hence meteorologists recognise both horizontal wind shear and vertical wind shear). Some element of wind shear is always present in the atmosphere. However, it becomes a significant hazard to aircraft when the change in wind velocity is very marked within a small height band in particular.

Figure M12.15

A marked windshear, through which a 20 knot headwind changes rapidly to a 10 knot tailwind.

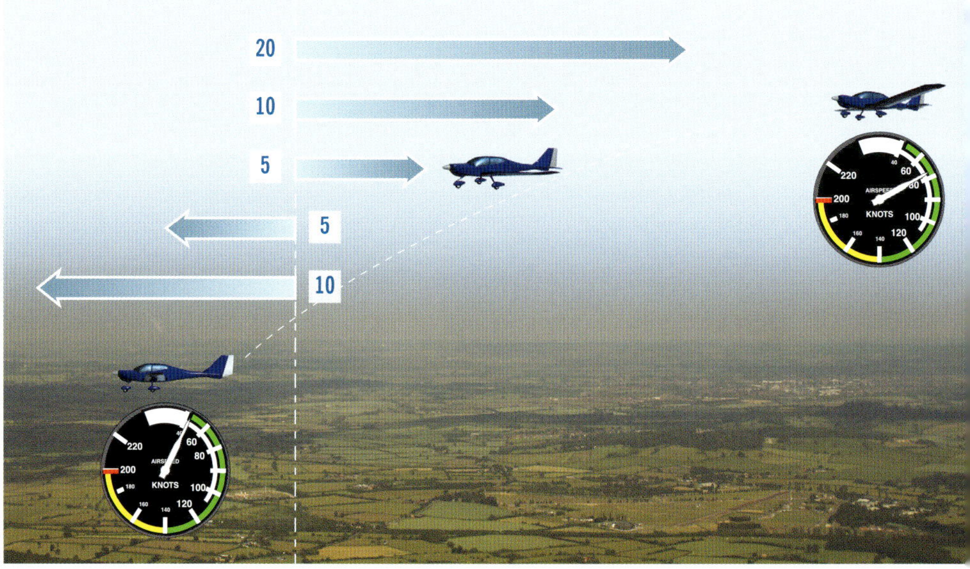

Figure M12.16

A strong low-level temperature inversion can lead to marked windshear through the top of the inversion layer.

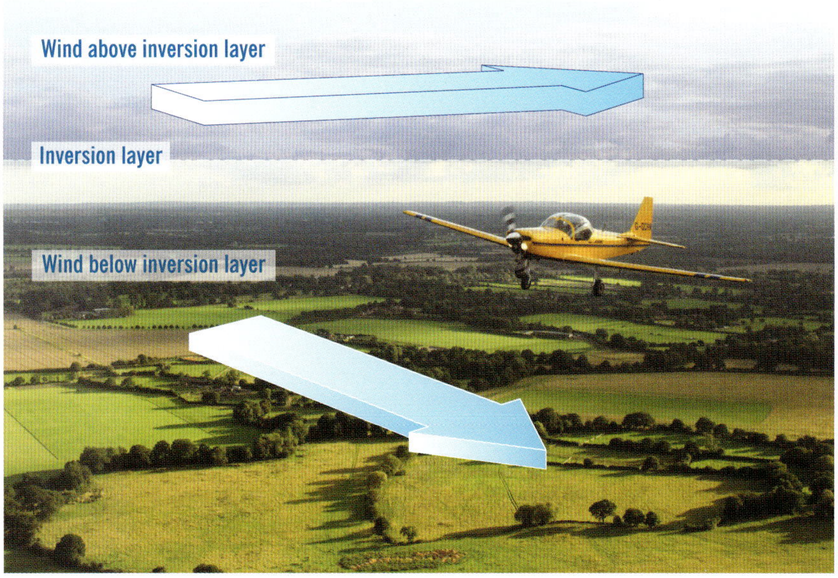

The real danger of a strong wind shear is the dramatic and rapid effect it can have on an aircraft's airspeed. Imagine the wind velocity changing so that an aircraft on approach to land transitions from experiencing a 20kts headwind component to a 10kts tailwind component within a 100ft height band. As it descends with an airspeed of 90kts, experiencing a 20kts headwind, the aircraft's groundspeed is 70kts. As the aircraft passes through the wind shear zone, the headwind of 20kts becomes a tailwind of 10kts. Due to inertia, the aircraft retains its original groundspeed for a short time, so the airspeed drops suddenly to 60kts. This naturally leads to a marked loss of lift, the aircraft will 'undershoot' the desired flight path and may even stall. At low level, recovery may be difficult or impossible.

Figure M12.17
Marked windshear is often present within and under a thunderstorm.

Wind shear in association with thunderstorms has already been discussed. Strong wind shear is also associated with strong winds at low level and can sometimes also develop when there is a marked temperature inversion near the surface. Such an inversion might exist on a clear night when air in contact with the surface is cooled, whereas the air higher in the atmosphere is warmer because it is not in contact with the surface. The result is to separate the friction layer from the general airflow, possibly leading to very strong winds at low altitude but almost calm conditions at the surface. Clear nights are a particular time to be wary of such low-level wind shear. Even if surface winds are calm or nearly so, the upper wind forecasts may indicate the possibility of much stronger winds above. In such circumstances, turbulence and wind shear may be very marked when passing through the transition zone – which can be as low as a few hundred feet above the surface.

If an aircraft experiences a 'negative' wind shear at low altitude (leading to a loss of airspeed and lift), the first priority is to retain control of the aircraft and if necessary use the stall recovery procedure. A go-around is likely to be prudent in such a situation.

In summary, low-level wind shear is most often associated with thunderstorms, marked temperature inversions and strong winds – especially where the terrain (such as a valley or ridge on the approach path) encourages its formation. Where wind shear is suspected or known to exist, the golden rule is avoidance. If this is not possible, approaching at a slightly higher than normal airspeed will give some element of protection, whilst remaining alert to the symptoms of wind shear, and being prepared to make an early decision to go-around.

Strong Winds

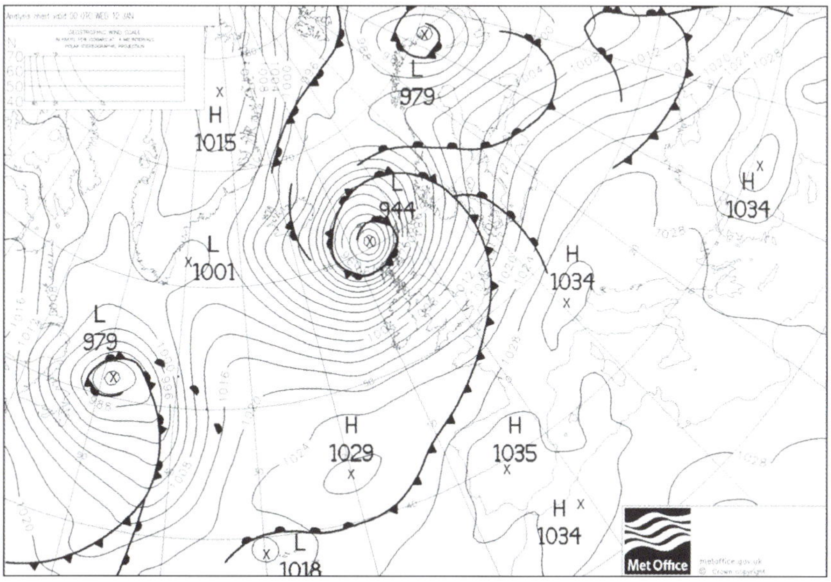

Figure M12.18
A deep depression in winter, passing just north of the UK.

Widespread **strong winds** are associated with a steep pressure gradient (marked on synoptic charts by closely packed isobars) – the steeper the pressure gradient, the stronger the wind. This makes the forecasting of gales and strong winds reasonably straightforward, and indeed it is comparatively rare for aircraft to be caught out by totally un-forecast strong winds.

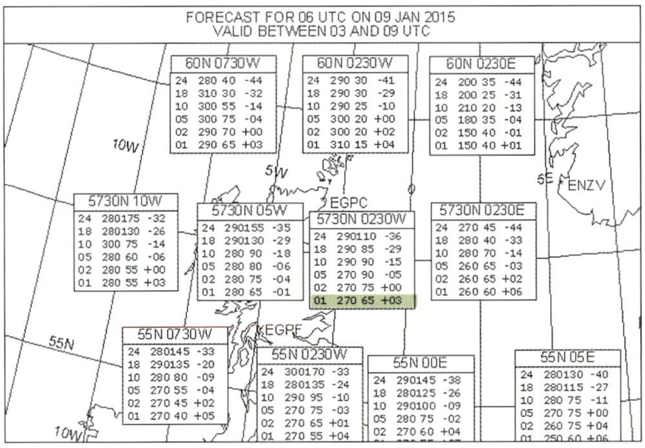

Figure M12.19
A 'spot wind' chart for the same day forecasting 65 knot winds at 1000ft in Scotland.

With the exception of flight over and near high ground, strong winds present the greatest danger to aircraft on and near the ground. The general advice to pilots is not to taxi, let alone take-off or land, if the surface wind speed is more than 50% of the aircraft's stalling speed. It follows that for most light aircraft, surface winds in excess of about 25kts demand extreme caution. Further, most light aircraft have much lower crosswind limits than this (often 12-15kts) so wind direction in relation to runway orientation is also an important consideration.

Figure M12.20
A 'TAF' aerodrome forecast for the day of the chart above. At Stornoway in the Scottish Hebrides the surface wind is forecast to be from 270° at 35 knots, but there is a 30% probability that temporarily between 06:00 and 08:00 the surface wind direction may become 260° and increase to 45 knots, gusting to 60 knots.

STORNOWAY EGPO 090502Z

0906/0915 27035KT 9999 FEW015 SCT030 PROB30
TEMPO 0906/0908 26045G60KT 6000 SHRA
BECMG 0909/0912 25018KT
PROB30 TEMPO 0911/0915 24021G31KT
BECMG 0912/0915 18018KT
TEMPO 0913/0915 4000 RA BKN008

Figure M12.21
A METAR aerodrome report for airfield EGPC (Wick, in northern Scotland).
On the 9th day of the month, at 05:50 UTC, the automated weather report has recorded a surface wind from 290 degrees, speed 57 knots gusting to 81 knots.

EGPC 090550Z AUTO **29057G81KT** 3400 HZ FEW016/// SCT026/// 05/02 Q0982

The strongest gales are usually associated with deep and fast-moving depressions, and 'lows' that are deepening rapidly. If the pressure at the centre of a low deepens at a rate of more than 24hPa in 24 hours, the winds are likely to be particularly strong. Such a fast deepening low is known to meteorologists as 'explosive cyclogenesis', or more commonly a weather 'bomb'. Of course, at any fixed spot the pressure may fall even more rapidly than this if the low is moving towards that place. However, the point is largely academic. If the pressure at a fixed location is falling rapidly (a rate of one hpa per hour or more), the probability is that bad weather and strong winds are imminent and appropriate precautions should be taken. When extreme lows do occur, they usually reach their maximum intensity over the ocean, at least as far as Europe is concerned. If an extreme low does cross the land while still at its lowest pressure, severe gales are likely.

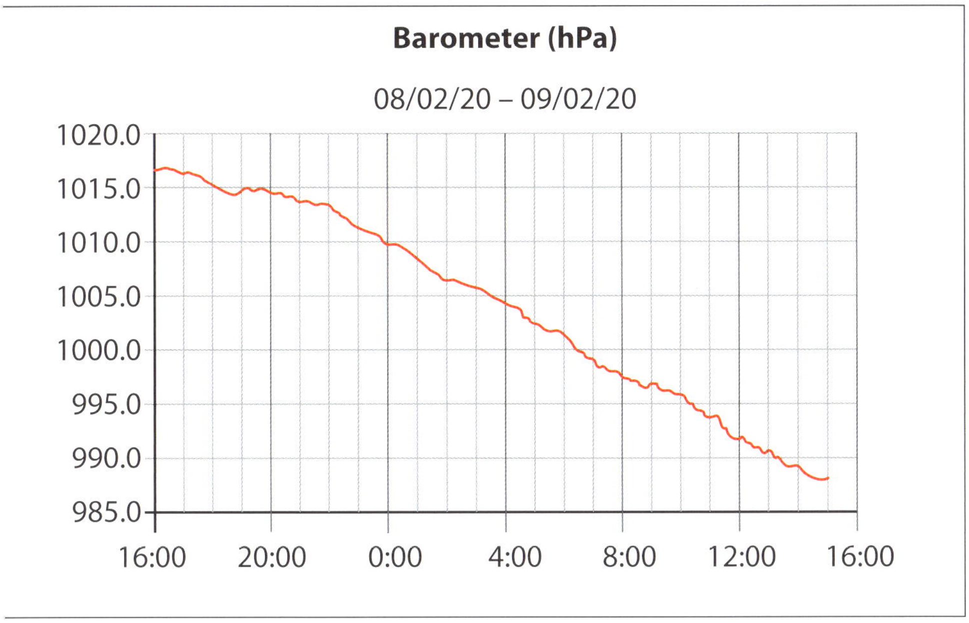

Figure M12.22
The trace of surface pressure at an airfield as a deepening depression approached. The rapid drop in surface pressure preceded strong gales.

When high winds are forecast, the best place for an aircraft is in a hangar. If this is not possible, the aircraft should be tied down in accordance with the instructions in the Aircraft Flight Manual (AFM). Merely parking an aircraft into wind and putting the control locks in will not guarantee adequate protection during a gale (defined as sustained surface winds of at least 34kts) or worse.

Finally, it is often assumed that the area downwind of high ground or an obstruction will be sheltered from the worst of the wind. This is not necessarily true. Downwind of high ground, wave activity can bring fast-moving air from altitude down to the surface. The result can be that surface winds downwind of hills or high ground are often more gusty and less predictable than elsewhere.

Figure M12.23
Aircraft parked outside need to be secured in accordance with the manufacturer's guidance when strong winds are expected.

Progress Check

73. Can a general airflow be modified in mountainous areas?
74. What is a 'katabatic' wind?
75. Is it necessary for cloud to be present for mountain waves to form?
76. Which hazardous situation maybe found just below and downwind of a ridge or mountain in strong wind conditions?
77. Where would you expect to find a 'rain shadow' is relation to a mountain range?
78. How is 'windshear' defined?
79. Which large scale meteorological situation is likely to lead to widespread gales?

These questions are intended to test knowledge and reinforce some of the key learning points from this section. In answering these questions, a 'pass rate' of around 80% should be the target.

Model answers are found at page M177.

M13 Meteorological Information

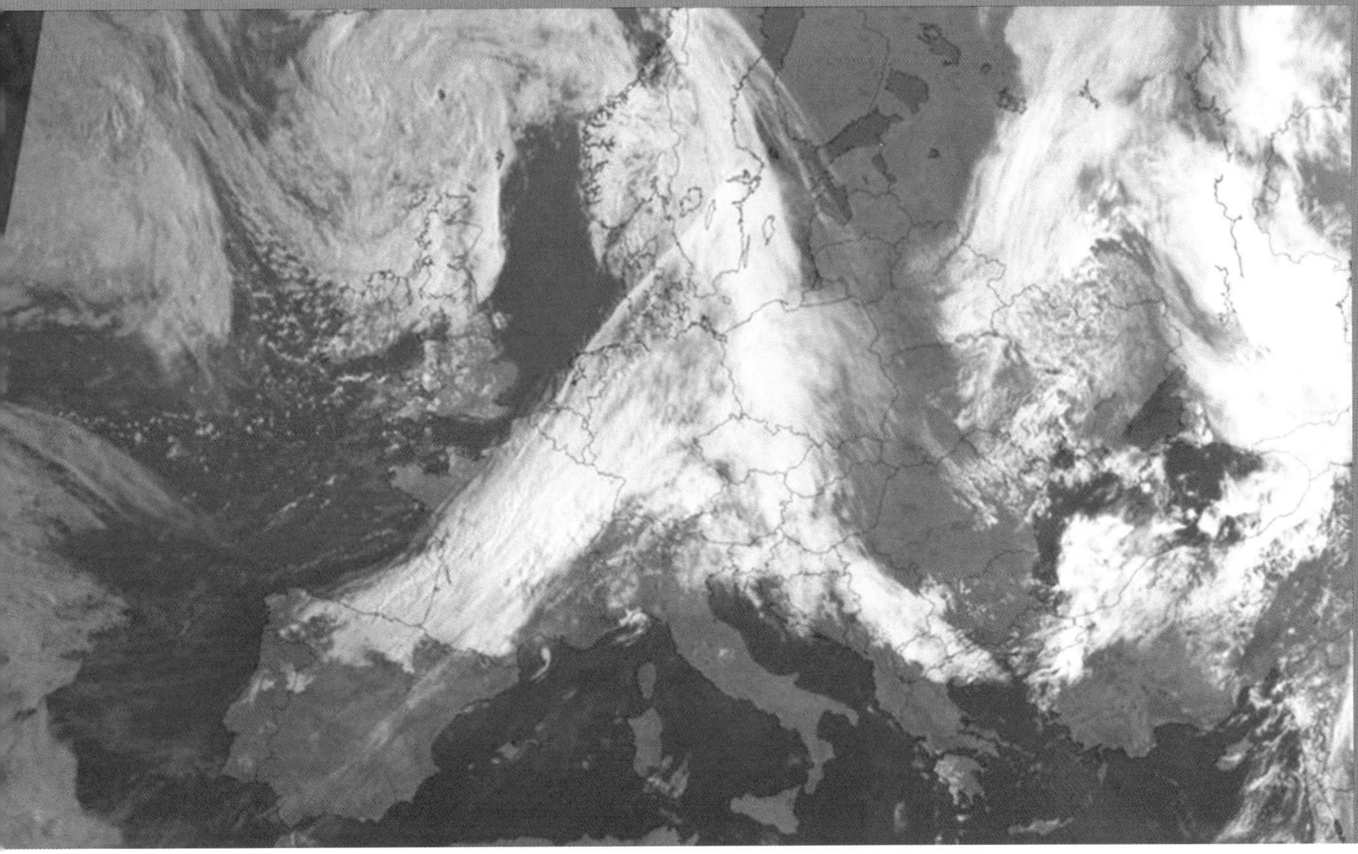

M13 Meteorological Information

Synoptic Charts

In meteorological terms, a **synoptic** chart is a chart giving a summary, a general view, of the key meteorological features over a large area (a horizontal distance of at least 1000km/620 miles) and typically covering an entire continent and/or ocean, or global region. The pressure distribution will be depicted using isobars, so that areas or high and low pressure are evident – these allow estimates to be made of wind direction and speed, as well as the general air mass affecting a particular area. Large scale weather features such as fronts (and maybe troughs) may also be portrayed, and again an interpretation of these features allows a judgement to be made about the sort of weather they will bring.

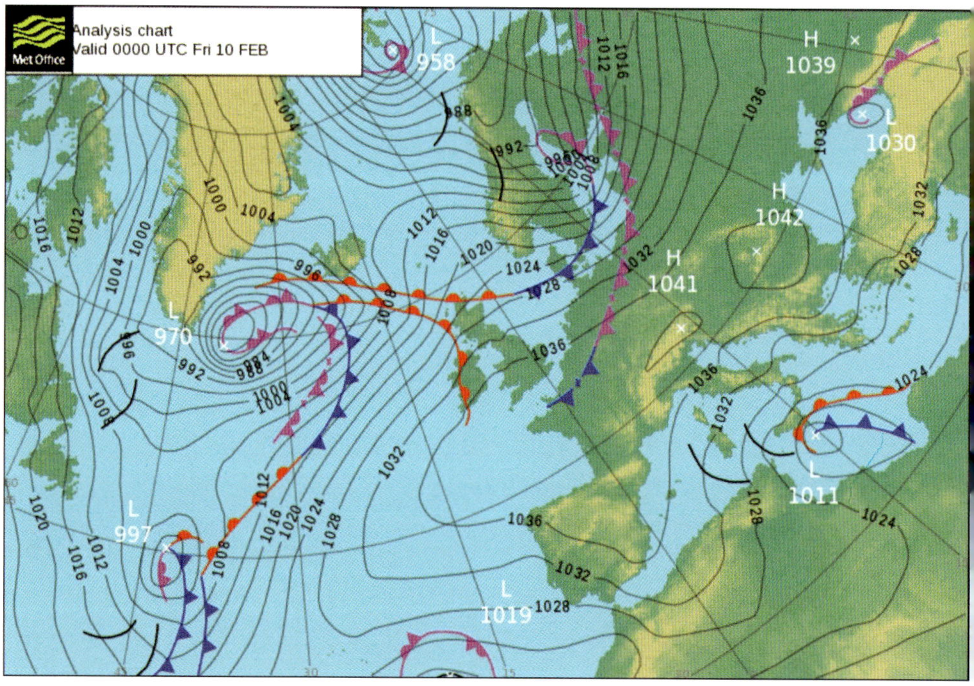

Figure M13.1
A synoptic chart shows the actual (observed analysis) or forecast weather situation over a large area, showing major weather features such as pressure distribution and fronts.

Synoptic charts can be of an observed situation, at a specific date and time, or in the form of forecast charts predicting the development and movement of the main meteorological features hours and days into the future. Synoptic charts (often with animated forecast charts) are widely used by broadcast weather forecasters and available on numerous websites.

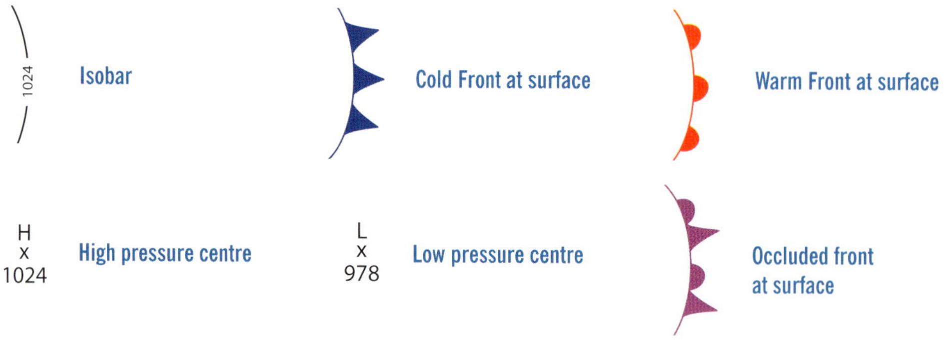

Figure M13.2
Symbols used on synoptic charts:

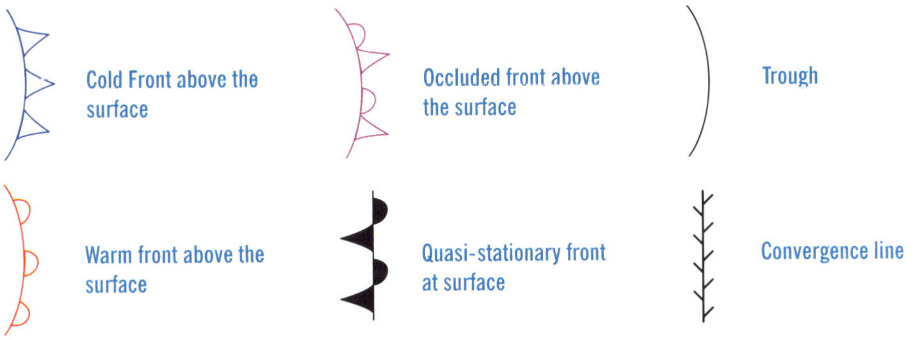

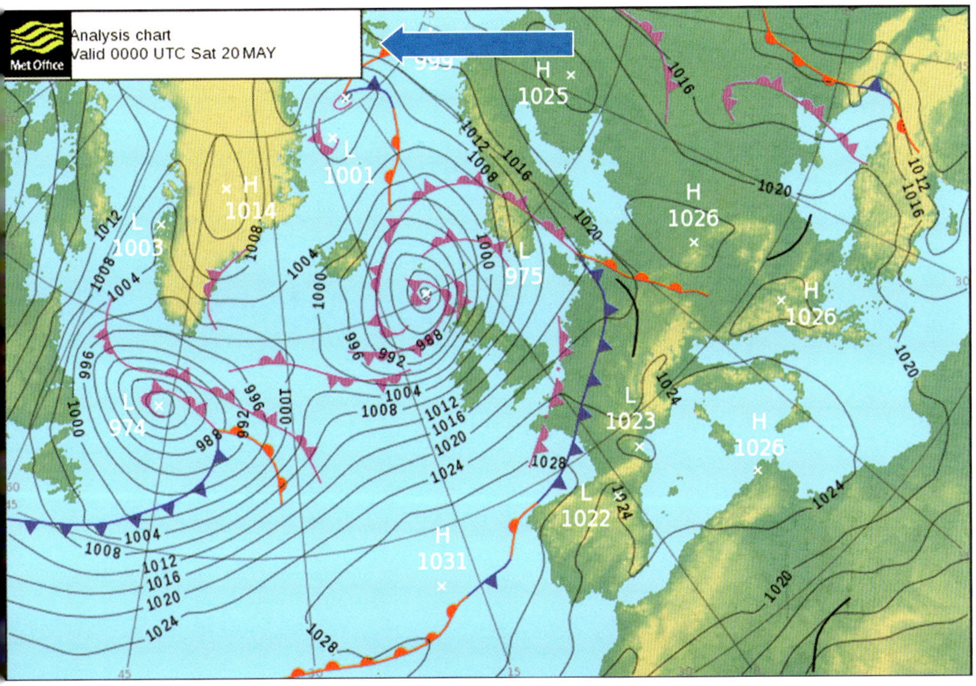

Figure M13.3

A UK Met Office 'actual' synoptic chart, the text box shows the time and date of the sea level observation; this is an analysis of the synoptic situation at that time.

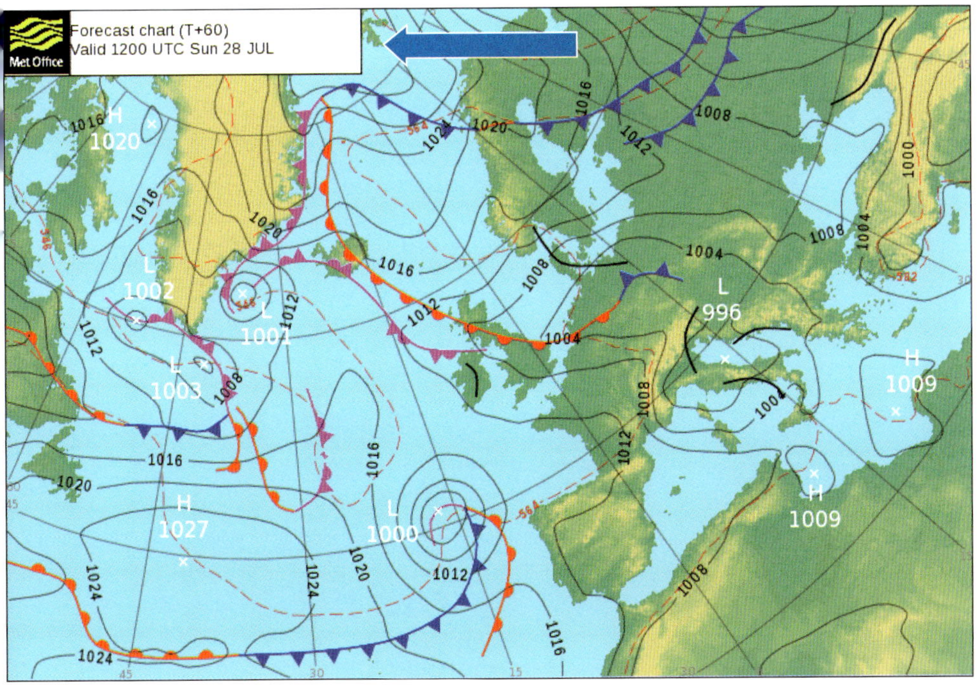

Figure M13.4

A UK Met Office forecast synoptic chart, the text box shows the time and date of the forecast situation, and that it is a forecast for 60 hours (T+60) ahead, ie two and a half days.

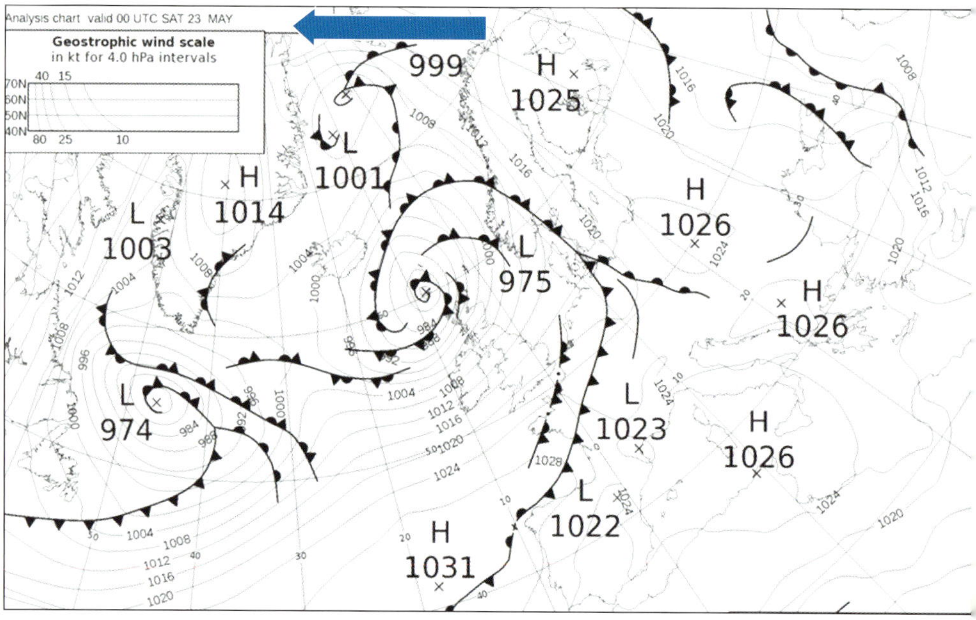

Figure M13.5

A black and white synoptic chart, the text box shows that this is an actual observation for the time and date noted.

Synoptic charts are usually based on the actual or forecast situation at **Mean Sea Level** (**MSL**), but to be sure of what the synoptic chart is depicting it is essential to check the chart legend for the 'what and when' information – it is simply not acceptable to make assumptions; the chart data must be checked, so that you know exactly what the chart is depicting and what date/time it relates to.

When preparing a weather forecast, a meteorologist starts with synoptic charts to get an overview of the current weather situation and how it is expected to develop. In much the same way, a pilot checking weather when planning a flight should start with the synoptic chart(s) to get the same summary overview of the general weather situation and how it is expected to develop.

Satellite Imagery

Whilst the use of satellites has revolutionised weather forecasting, the practical use of satellite imagery by pilots has more limited benefits. Nevertheless, satellite imagery is now widely available via the Internet and is available in two basic formats – visible image and infrared (IR).

Figure M13.6

A visible satellite image of Europe.

Figure M13.7
An Infrared satellite image of Europe.

A **visible satellite image** is to all intents and purposes a photograph taken from the satellite. Thick cloud tends to show as white, and coastlines and snow-covered ground are also clearly visible where cloud cover permits. Unfortunately, ground covered with snow may be confused with thick cloud, and thin cloud (such as cirrus) may be almost transparent and hence difficult to see. Additionally, a satellite cannot produce a visible image at night. The black (or blanked-out) area on a satellite picture taken near sunrise or sunset is an area in darkness. However, areas where the sun is low in the sky can give good images of cloud shadows which can give some indication of the vertical extent on cumulus-type clouds. The resolution of imagery available to general users will vary, but many websites allow an animation of recent images to be run, so that the movement and development of areas of cloud can be tracked.

An **infrared (IR) satellite image** is based on temperature differences. As such, it is not influenced by the absence of daylight. Cold areas generally show up as white on an IR picture, with darker shades of grey for warmer areas and black for those which are warmest. This type of imagery is useful but has certain limitations. High cloud, being coldest, tends to show as a bright white area. Therefore, even a very thin layer of high cirrus might totally obscure anything below it. Closer to the surface, very low cloud or fog may have nearly the same temperature as the surface, making it very difficult to distinguish

Satellite images are subjected to a certain amount of processing before they reach the pilot or forecaster. Latitude and longitude grid lines may be added, as well as coastlines, and the image may be adjusted for the amount of daylight depending on the time of day. Even so, satellite images are generally considered to be of little use in isolation, and in pre-flight planning they should only be considered in conjunction with other synoptic or significant weather information.

Ground-based Weather Radar

Most European nations have a network of **weather radar** stations that provide information on precipitation. The principle of radar is that the radar station transmits a series of radar waves, some of which reflect (echo) back to the station. More dense objects (such as large water droplets) provide a stronger return (echo) than smaller, less dense objects. With the aid of considerable processing power and clever software, the data from a radar station (or network of radar stations) can be processed into a pictorial presentation of how heavy the precipitation is and where it is falling. By viewing an animation of recent radar images, meteorologists can also make an informed estimate of where precipitation is likely to occur in the next few hours.

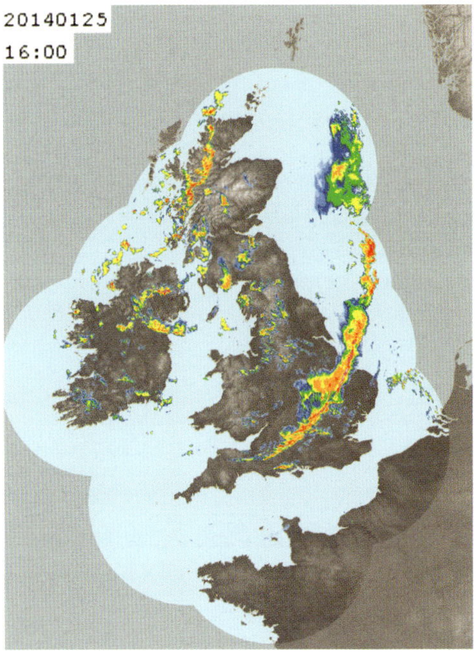

Figure M13.8
A weather radar image of the UK, showing areas of intense rainfall, most likely (in this case) to be thunderstorms.

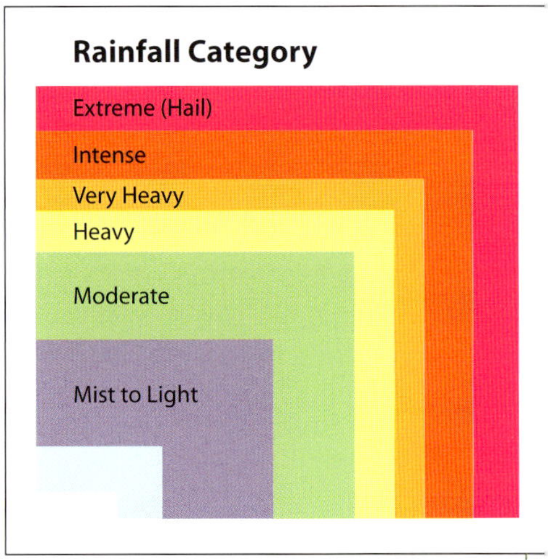

Figure M13.8a
Colour coding on a weather radar image.

The normal convention is that light precipitation is represented by light blue colouring, progressing through increasingly bright colours (amber and yellow), with heavy precipitation being marked by vivid red and purple colours. A weather radar image often has a colour scale which shows the colours used to represent the approximate rate of precipitation.

Like other forms of radar, weather radar has some limitations. If high ground blocks the radar signals, there may be an arc of the coverage where there is no data. Additionally, the fact that radar waves (usually) travel in straight lines means that, due to the curvature of the earth, precipitation at low levels at the far edge of the radar's range may not be visible to the radar station. It is also important to remember that cloud and fog are invisible to most radar systems, indeed very light precipitation (such as drizzle) may also consist of water droplets too small to be picked by radar. Therefore, very nasty flying conditions – such as fog, low cloud and drizzle – may be occurring in areas which appear 'clear' to weather radar.

Notwithstanding these limitations, weather radar can be very useful in tracking the movement of large scale weather systems such as depressions, fronts, etc. as well as more localised phenomena such as showers and thunderstorms.

Weather radar images are widely available via the Internet although they are often delayed by at least 15 minutes rather than being real-time. As with all other weather information, it is essential to check the actual date and time of the observations being viewed – it may be necessary to 'refresh' web pages to get the most up-to-date information available.

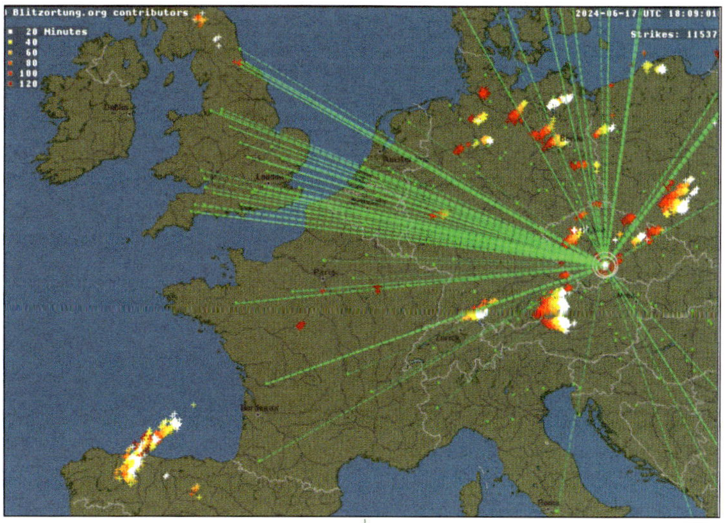

Incidentally, modern Air Traffic Control (ATC) radar is specifically designed to filter out weather 'clutter' from its screens, so in most circumstances radar controllers are unable to offer much help from their radar as far as weather avoidance is concerned.

Many European countries also have ground-based lightning detectors which pinpoint active thunderstorms, based on their electrical activity. This system assists in forecasting and can also provide useful real-time information to a pilot on a day when thunderstorms are forecast by displaying the intensity of lightning strikes. Often lightning strikes are colour coded by how long ago they occurred, to enable the movement of thunderstorms to be tracked.

Figure M13.8b

A thunderstorm tracking website, showing extensive lightning strikes in central Europe.

Area and Significant Weather Forecasts

Many nations with significant aviation activity produce aviation-specific weather forecasts, usually covering an area somewhat smaller than most synoptic charts.

In some European states, area forecasts specifically designed for VFR pilots may use a combination of colour coding and letters to indicate weather conditions in specified areas and/or routes. As the presentation and criteria of these coding varies from country to country, it is all the more important to study the notes on the forecast carefully so that you are absolutely sure what the forecast is predicting. This is in addition to the all-important checking of the date and time period of any forecast.

Aside from any forecast-specific coding most area or significant weather forecasts will tend to use established abbreviations and/or symbols to indicate cloud, weather, wind etc. Vertical distances are given as 'altitudes' (that is, Above Mean Sea Level –AMSL) and wind directions in degrees 'true', unless stated otherwise.

Figure M13.8c

Many European states produce some form of area forecast for General Aviation operations; this chart is for the Channel islands area.

FORECAST FOR FLIGHTS BELOW 15,000FT WITHIN A 100KM RADIUS OF JERSEY **01 JUNE 2024 : VALID 1600 TO 2200 UTC**

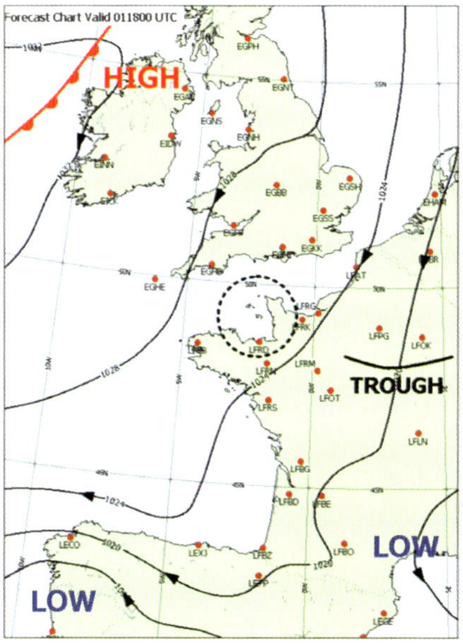

GENERAL SITUATION AT 1200 UTC

A northeasterly flow covered the area.

WINDS/TEMPS	**POINT**	**ORTAC**		**LFRD**	
	1000 FT	**030/10KT**	**PS14**	**010/15KT**	**PS12**
	2000 FT	**030/20KT**	**PS12**	**030/15KT**	**PS10**
Valid at 1900 UTC	**5000 FT**	**020/20KT**	**PS05**	**030/20KT**	**PS07**
	10000 FT	**020/20KT**	**MS01**	**030/15KT**	**MS02**
	18000 FT	**020/25KT**	**MS16**	**020/25KT**	**MS17**

FREEZING LEVEL 9500FT

AIRFRAME ICING LCA MOD

WEATHER Local haze, mainly in France.

VISIBILITY 20km or more. LCA 6-10km in HZ.

CLOUD 1-3/8 BUT IN E 4-7/8 CU/SC 1500-4000FT/3500-7000 OR MERGING
OCNL IN E 1-4/8 AC 7000-9000FT/8000-12000

WARNINGS LCA MOD

F/C LOWEST MSL PRESSURE 1024 **hPa**

OUTLOOK TO 02 JUNE 0900 UTC

Local haze and a risk of mist patches inland France from 0000 UTC.

Note: ALL HEIGHTS ARE AMSL AND CLOUD IN OKTAS. CLOUD HEIGHTS ARE BASE/TOPS: TWO HEIGHTS SEPARATED BY HYPHEN(-) SHOW POSSIBLE VARIATION IN BASE OR TOPS.

ISSUED BY JERSEY MET AT 14:51 UTC
FORECASTER MW © Government of Jersey 2024

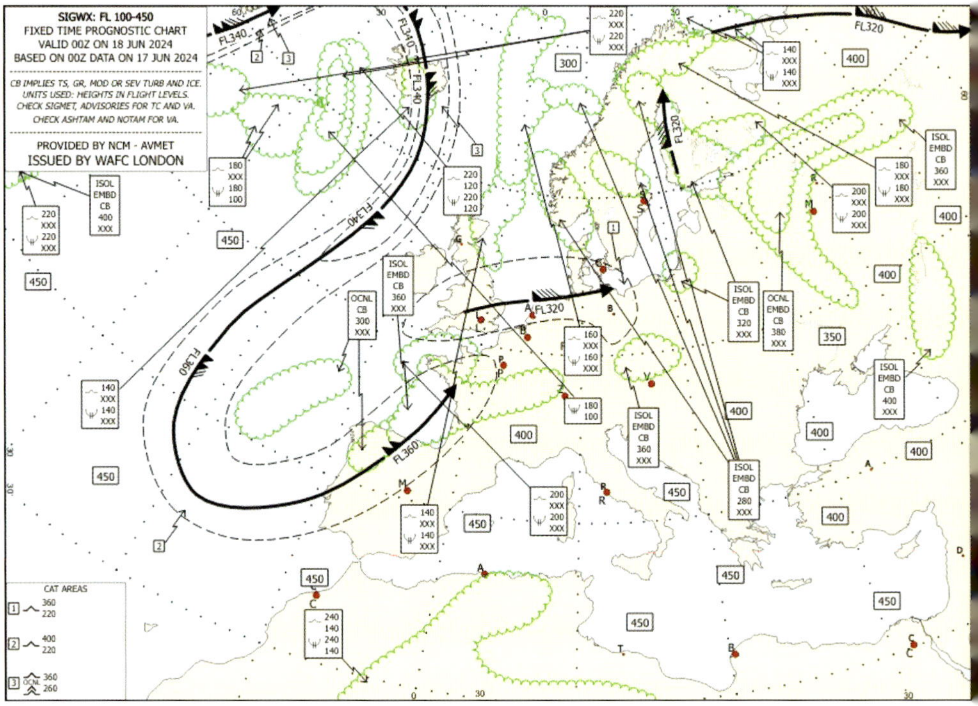

Figure M13.8d

A European SIGWX chart for the levels between Flight Level 100 and Flight Level 450.

A **Significant Weather Chart** (sometimes known as **SIGWX**) displays observed or forecast significant weather phenomena that may affect the operation of aircraft. A Significant Weather Chart covering higher flight levels is usually restricted to showing:

- Areas of cumulonimbus cloud activity.
- Areas of moderate or severe icing.
- Areas of moderate or severe turbulence.
- The position of jet streams and any associated clear air turbulence (CAT).
- The flight level at which the tropopause can be found.
- Volcanic ash warnings.

The following symbols are commonly used in high level SIGWX charts:

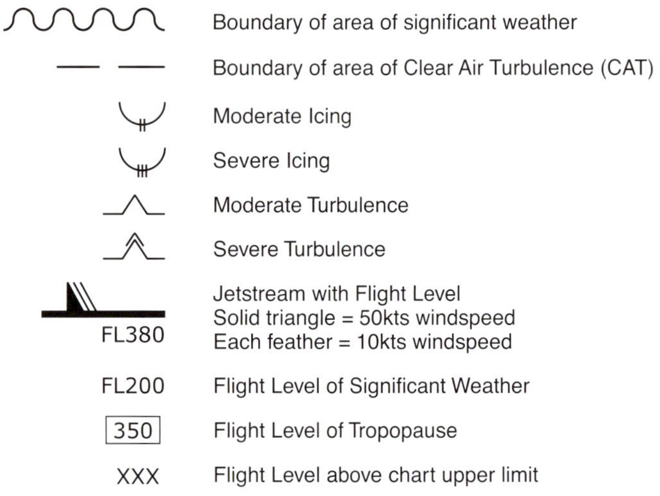

Figure M13.9

Common SIGWX symbols.

The following abbreviations may be used in high level SIGWX charts:

BLW	Below	GEN	Generally	NC	No Change
CAT	Clear Air Turbulence	HVR	Heavy	OCNL	Occasional
CB	Cumulonimbus Cloud	ISOL	Isolated	SEV	Severe
CLD	Cloud	LAN	Land	STNR	Stationary
COT	Coast	LCA	Local / Locally	TURB	Turbulence
EMBD	Embedded in other cloud	LYR	Layer	VSP	Vertical Speed
FCST	Forecast	MOD	Moderate	WDPR	Widespread
FPM	Feet Per Minute	MOV	Moving	WRNG	Warning
FRQ	Frequent	MTW	Mountain Wave		

Figure M13.9a
Common SIGWX and Area Forecast abbreviations:

Many countries produce forecasts of upper winds and temperatures at stated levels through the atmosphere.

TAFs and METARs

TAFs (**T**erminal **A**erodrome **F**orecast) and METARs (**Met**eorological **A**erodrome **R**eport) are forecasts and actual observed weather at individual airfields. Although at first sight they can appear to be a heavily coded series of random letters and numbers, with a little practice they are surprisingly intuitive to understand. They are also highly standardised worldwide so if you can read and interpret a METAR or TAF for London Heathrow, for example, you will also be able to understand the salient points of a METAR or TAF produced by almost any European state, and indeed most states worldwide (although some countries use non-standard units of measurement for items such as visibility and temperature, for example).

A **METAR** (defined as an 'aerodrome routine meteorological report') is just that, it is a report of the actual weather conditions observed at a specific place (an airfield) and at a specific time. A METAR follows a prescribed format and most of the codes and conventions used in a METAR are also used in a TAF.

When reading any weather observation, it is vital to first establish the 'where and when' – without this vital information, everything that follows is worthless. A METAR will contain some or all of the following data, always reported in this specific order:

Location | Date & Time | Surface Wind | Visibility | Present Weather | Cloud | Temperature and Dew Point | QNH | Recent Weather | Trend Forecast

LSGG 070820Z VRB03KT 3500 BR OVC003 07/06 Q1029 RERA NOSIG

Figure M13.10

Location	Date & Time	Surface Wind	Visibility	Present Weather	Cloud	Temperature and Dew Point	QNH	Recent Weather	Trend Forecast
LSGG	070820Z	VRB03KT	3500	BR	OVC003	07/06	Q1029	RERA	NOSIG

LOCATION: In a METAR, the identification of the reporting station is indicated using the airfield's four letter ICAO location indicator. These location indicators can be checked in any number of aeronautical publications and flight guides. The first two letters of the ICAO location indicator are a country code (for example 'ED' for Germany, 'EG' for the United Kingdom; 'LF' for France, etc.). The second two letters are allocated to a specific airfield in that country. For example, EGLL is London Heathrow; LFPG is Paris Charles de Gaulle and EDDF is Frankfurt Main. In practice, you will soon get to know the ICAO location indicators for those airfields whose METARs you check regularly, and often on-line METAR reports are formatted to include the airfield name as well as the ICAO location indicator. A METAR will report conditions at the airfield, and within approximately 8 kilometres (5 nautical miles) of the centre of the airfield.

DATE and TIME: The date and time of the METAR observation is given in a six digit group: the first two digits are the date, followed by a four digit group for the time of the observation in hours and minutes in UTC (also known as 'Zulu' time). To confirm this, the date/time group is followed by the letter 'Z' to denote that the observation time is in UTC. For example: '020820Z' is an observation on the second day of the month ('02') at a time of 0820 UTC.

AUTOMATED REPORT: A fully automated METAR observation is indicated by the word 'AUTO'.

SURFACE WIND: The direction and strength of the surface wind (representative of the wind at a height of approximately 10 meters / 30 feet affecting the runway/runways) is reported using degrees true for direction (that is, referenced from True north) rounded to the nearest 10 degrees, and (usually) knots for wind strength, to the nearest knot. The wind velocity (w/v) reported is an average of that observed in the previous 10 minutes before the time of the METAR. For example: '320/10KT' indicates a mean surface wind in the 10 minutes before the METAR observation time with a direction of 320 degrees true, at 10 knots (**KT**).

If there is a significant variation in wind strength, the letter '**G**' will be inserted ('G' for 'Gusting'), followed by the maximum wind gust in the previous 10 minutes. For example: '07021G32KT' is a report of a surface wind with a direction of 070 degrees (true), with a '10 minute' mean speed of 21 knots, gusting to a maximum of 32 knots in that period.

If the wind direction varies by 180 degrees or more over the 10 minute observation period, the direction digits will be replaced by '**VRB**' – variable. For example: 'VRB05KT' is a surface wind varying in direction by more than 180 degrees, at 5 knots.

If the wind direction varies by up to 180 degrees, an extra set of digits will be reported after the mean wind velocity representing the extremes of the wind direction variation, separated by '**V**' for 'varying'. For example: '08012KT 050V110' is reporting a surface wind with a mean direction of 080 degrees at 12 knots, but with the wind direction varying between 050 degrees and 110 degrees.

If the wind strength is less than 1 knot, direction and speed are replaced with '**00000**', spoken as "wind calm".

Although most European states report wind strength in 'knots', you may occasionally see wind speed reported in metres per second – 'MPS'. 1MPS is equivalent to approximately 2 knots.

VISIBILITY: In European states, observed visibility is given as a four-digit number, reported in metres. Between 800 metres and 5 kilometres, visibility is reported in 100 metre steps; between 5 and 10 kilometres, visibility is reported to the nearest kilometre. If visibility is more than 10 kilometres, it is reported as '**9999**', often spoken as "*all the nines*" or "*ten plus*".

The visibility reported in a METAR will be the 'prevailing visibility' – pilots do not need to know the method of calculating prevailing visibility, but you should be aware that the visibility you experience looking out from an aircraft may be quite different (better or worse) than the 'prevailing visibility' measured on the airfield. For example: a reported visibility of '8000' represents a prevailing visibility of 8000 metres or 8 kilometres.

If there is particularly notable variation in visibility when looking in different directions, the prevailing visibility is given followed by the minimum visibility and the direction in which that visibility is observed in relation to the airfield, using the 'cardinal' points of the compass:

N North
NE North East
E East
SE South East
S South
SW South West
W West
NW North West

For example: '9999 6000E' indicates a prevailing visibility of more than 10 kilometres, but a minimum observed visibility of 6000 metres looking to the east of the airfield.

Where visibility has been measured by an automated sensor, the reported visibility value may be followed by '**NDV**', indicating that No Directional Variation in visibility is available.

If visibility is less than 1500 metres, in addition to prevailing visibility, Runway Visual Range ('RVR') specific to a stated runway may be reported. For example: 'R26/0600' is an RVR along runway 26 of 600 metres. As such visibilities are well below most VMC minima, RVR is not explored further here; however pilots studying for some form of instrument flying qualification can look forward to learning much more about the measurement and reporting of RVR.

PRESENT WEATHER: Any 'weather' observed at the airfield will be reported using the following abbreviations (some of which are sufficiently obscure that they may challenge even experienced pilots). They are grouped together here by different types of weather, up to three types of weather may be reported in a METAR (in which case the conditions are likely to be particularly interesting).

Phenomena obscuring visibility:

Abbreviation	Meaning	Comments
BR	Mist	Used when visibility is between 5km and 1000m
DU	Dust	
FG	Fog	Generally used when visibility is less than 1000m
FU	Smoke	
HZ	Haze	
SA	Sand	
VA	Volcanic Ash	

Precipitation:

Abbreviation	Meaning	Comments
DZ	Drizzle	
GR	Hail	(largest hailstones 5mm diameter or more)
GS	Small hail /snow pellets	(largest hailstones less than 5mm diameter)
PL	Ice Pellets	
RA	Rain	
SG	Snow Grains	
SN	Snow	
UP	Unknown Precipitation (only used in automated METARs)	

Weather:

Abbreviation	Meaning	Comments
BC	Patches	Random fog patches covering the airfield
BL	Low Blowing	DU, SA or SN raised by the wind to 2m (7ft) or more above the surface
DR	Low Drifting	DU, SA or SN raised by the wind to less than 2m (7ft) above the surface
FZ	Freezing	
MI	Shallow	Less than 2m (7ft) above the surface
PR	Partial	A substantial part of the airfield is covered by fog
SH	Shower	
TS	Thunderstorm	

Other weather phenomena:

Abbreviation	Meaning	Comments
DS	Dust Storm	
FC	Funnel Cloud	Associated with a tornado or waterspout
PO	Dust devils/ sand whirls	
SQ	Squall	
SS	Sand Storm	

Intensity:

Abbreviation	Meaning	Comments
+	Heavy	
[no symbol]	Moderate	
-	Light	

Proximity:

Abbreviation	Meaning	Comments
VC	Vicinity	Between approximately 8km and 16km from the centre of the airfield

For example: 'MIFG' indicates shallow fog; '-SHSN' indicates light showers of snow; 'VCSH' indicates showers in the vicinity (between approximately 8km and 16km of the airfield centre); 'RA BR' indicates moderate rain and mist.

CLOUD: Any significant cloud within approximately 16km of the centre of the airfield is reported, the height (above aerodrome level) of the cloud base is reported in units of 100ft. The amount of cloud is measured in terms of 'oktas' (an okta being 1/8th of the sky) using the following abbreviations:

FEW	Few	1-2 oktas
SCT	Scattered	3-4 oktas
BKN	Broken	5-7 oktas
OVC	Overcast	8 oktas

For example: 'FEW015 SCT030' is a report of 1-2 oktas of cloud with a base of 1500ft above the airfield, and additionally 3-4 oktas of cloud with a base of 3000ft above the airfield.

Cloud type is not normally given in a METAR, except:

CB Cumulonimbus

TCU Towering Cumulus

For example: 'BKN014CB' is a report of 5-7 oktas of Cumulonimbus cloud with a base of 1400ft above the airfield.

In some conditions, Vertical Visibility ('**VV**') may be reported, followed by a vertical distance in feet. For example: 'VV002' is reporting a vertical visibility of 200ft above the airfield. '**000**' indicates a cloud base of less than 100ft above the airfield.

NCD indicates that in an automated METAR, No Cloud Detected.

NSC indicates Nil Significant Cloud

SKC indicates SKy Clear

The visibility and cloud reports may be replaced with the term '**CAVOK**' (meaning Ceiling And Visibility OK and spoken as "*CAV-OK*") when the following conditions are observed:

- Visibility – 10km or more;
- Cloud – no cloud below 5000ft above the airfield, or no cloud below the highest Minimum Sector Altitude, whichever is the higher;
- Cloud type – no Cumulonimbus cloud and no Towering Cloud;
- Weather – no significant weather phenomena.

EMPERATURE and DEW POINT: Surface air temperature and dew-point are normally given in whole degrees Celsius (centigrade). For example: '20/09' indicates a temperature of +20 degrees Celsius, and a dew point of +9 degrees Celsius.

A temperature value below freezing (0°C) is preceded by '**M**', for 'Minus'. For example: '01/M03' indicates an observed air temperature of +1°C, with a dew point of -3°C

QNH: The airfield QNH is reported in hectopascals as a four digit number, preceded by '**Q**'. For example: 'Q1026' is an observed airfield QNH of 1026 hectopascals.

RECENT WEATHER: A METAR may include a report of recent weather observed at the airport since the last report, or in the last hour, whichever is shorter (but not occurring at the time of the observation). Recent weather will be indicated by the abbreviation '**RE**' followed by the abbreviation for the weather phenomena. For example: 'RERA' indicates REcent RAin.

TREND: At some airfields, a 'TREND' forecast may be added to the METAR observation. The TREND information is a forecast of any significant changes in meteorological conditions expected in the next two hours (that is, two hours from the time of the METAR observation). In addition to the standard abbreviations already described, a TREND forecast may use one or more of the following abbreviations:

TEMPO	(Temporarily) forecast temporary fluctuations to meteorological conditions expected to last for a period of less than one hour in each instance and overall cover less than half of the forecast period.
BECMG	(Becoming) used to describe expected changes to meteorological conditions at either a regular or irregular rate.
FM	(From), followed by a four digit time group in hours and minutes UTC.
TL	(Until), followed by a four digit time group in hours and minutes UTC.
AT	(At), followed by a four digit time group in hours and minutes UTC.
NOSIG	(No Significant Change), used when none of the conditions in the METAR observation are expected to change significantly.
NSW	Nil Significant Weather

For example: 'BECMG 9999 NSW SCT015' is a trend forecast of conditions in the two hours after the observation time of visibility becoming 10 kilometres or more, Nil Significant Weather, cloud base scattered at 1500ft above the airfield.

RMK (Remarks) denotes the beginning of a section containing information included by national decision, which may not be distributed or recognised internationally.

// or /// A series of 'slashes' indicates that data is not available for that section of the METAR. For example: 'FEW014///' indicates cloud base few at 1400ft above the airfield, cloud type unavailable. These symbols are most often found in automated METARs.

= Indicates the end of the report.

Very occasionally a special weather report (observation) may be issued when there is significant deterioration or improvement in airport weather conditions between scheduled METAR observations. A special weather report is preceded by the term **SPECI**, otherwise the standard METAR terms and codes are used.

A **TAF** (Terminal Aerodrome Forecast) is a prediction, an estimate, of conditions that are expected to occur during a specified time period. It is important to understand the fundamental difference between a METAR observation, which is an accurate report of conditions at a specific time, and a TAF forecast of conditions expected to occur up to 36 hours in the future. Some pilots are known to describe a TAF as 'a horoscope with numbers' – which may be deeply unfair to the professionalism of weather forecasters (and their computers), but does put across the point that TAFs do not come with any form of guarantee!

A TAF will contain some or all of the following predicted conditions, always forecast in this specific order:

Location | Date/time at which the forecast was prepared | Validity period | Surface Wind | Visibility | Significant Weather | Cloud | Significant Changes

Figure M13.11

EGTK 251653Z 2518/2522 14009KT 5000 HZ FEW045 PROB30 TEMPO 2518/2520 4000 SHRA TSRA PROB30 TEMPO 2520/2522 3000 +SHRA +TSGS

Location	Date/time at which the forecast was prepared	Validity period	Surface Wind	Visibility	Significant Weather	Cloud	Significant Changes
EGTK	251653Z	2518/2522	14009KT	5000	HZ	FEW045	TEMPO 2518/2520 4000 SHRA TSRA PROB30 TEMPO 2520/2522 3000 +SHRA +TSGS

For the most part, a TAF forecast uses the same abbreviation and protocols as a METAR report, so the following only details those factors which are unique to a TAF.

TIME/DATE AT WHICH THE FORECAST WAS PREPARED Not always shown, this is six-digit group of the date (first two digits) and time in hours and minutes (next four digits) of the origin of the forecast – the date and time at which it was prepared.

For example, '040503Z' indicates that the forecast was produced on the 4th day of the month at 0503 UTC.

VALIDITY PERIOD The dates and times of the start and end of the forecast period, as four digit groups – that is date (a two digit date) followed by a time (two digit hours in UTC). For example: '0506/0515' is a forecast validity period from the 5th at 0600 UTC to the 5th at 1500 UTC; a figure of '1306/1412' is a forecast validity period from the 13th at 0600 UTC to the 14th at 1200 UTC.

SIGNIFICANT CHANGES Dates and times during which significant changes may occur are given using four digit date/time groups. For example: 'TEMPO 0609/0612' is a forecast temporary change in weather conditions between 0900 UTC on the 6th and 1200 UTC on the 6th.

Forecast significant changes may come with a probability factor, a percentage of either 30% or 40% will be given.

PROB30 A 30% probability of the significant change occurring.

PROB40 A 40% probability of the significant change occurring.

For example: 'PROB30 TEMPO 0506/0510 7000 -DZ SCT004' is a forecast of a 30% probability of a significant change in conditions, namely that in the period 0600 UTC on the 5th to 1000 UTC on the 5th, conditions will temporarily change to a visibility of 7000 metres, light drizzle and cloud base scattered at 400ft above the airfield.

Other abbreviations The following abbreviations may accompany or replace a TAF:

AMD Amended

CNL Cancelled

COR Corrected

NIL Missing forecast

his description of common METAR and TAF codes and abbreviations is intended to cover ne formats commonly found in METARs and TAFs for most European airfields. To cover very possible permutation that may be used anywhere in the world would need a book in self (the World Meteorological Organisation manual on codes alone runs to 480 pages!). herefore, if you come across a code or abbreviation you do not recognise, it pays to find ut what that figure or abbreviation actually means. Most aviation forecasting offices will ave a website which includes the abbreviations they use, most national Aeronautical nformation Publications (AIP) and commercially produced flight guides also have extensive ables of METAR and TAF decodes, including some decidedly uncommon codes. As in all natters of aviation safety, if you do not understand something, find out or ask someone!

On first reading, the codes and abbreviations used in METARs and TAFs can occasionally ppear to be baffling. The best way to become comfortable in understanding METARs and AFs is simply to read them as often as possible. As described shortly, METARs and TAFs are videly (and freely) available, not least online. Reading just one METAR and TAF a day will oon give you confidence in understanding them and also contribute to a deeper ppreciation of how wider weather trends translate into local conditions.

ources of Meteorological Information

viation legislation is very specific in placing a legal duty on the pilot to check before flight nat the weather conditions are suitable for the planned flight. For example, this is an xtract from the 'Part NCO' regulation:

"Before commencing a flight, the pilot-in-command shall be familiar with all available meteorological information appropriate to the intended flight. Preparation for a flight away from the vicinity of the place of departure, and for every flight under IFR, shall include:

(1) a study of available current weather reports and forecasts; and

(2) the planning of an alternative course of action to provide for the eventuality that the flight cannot be completed as planned, because of weather conditions."

Most professional pilots will agree that even with 1000s of flights in their logbook, each ight is different and very often it is the weather that makes that difference, Actual and

1.1.2 ADMINISTRATION METEOROLOGIQUE

METEO-FRANCE est l'autorité administrative météorologique désignée par l'état français pour la fourniture de l'assistance météorologique à la navigation aérienne.

1.1.2 METEOROLOGICAL ADMINISTRATION

METEO FRANCE is the meteorological administrative authority appointed by the French state to provide meteorological assistance to air navigation.

	Adresses postales / Postal addresses :	
METEO FRANCE	Aéroport de CAYENNE FELIX EBOUE	METEO FRANCE
73, Avenue de Paris	Service météorologique	Aérodrome de Saint Pierre
94165 SAINT MANDE CEDEX	97351 MATOURY	BP 4232 Saint Pierre
TEL : 01 77 94 77 94	TEL : 05 94 35 35 37	97500 SAINT PIERRE ET MIQUELON
FAX : 01 77 94 70 05		TEL : 05 08 41 18 60
TELEX : METFRAN 202 876		FAX : 05 08 41 18 65
		E-mail : meteospm@cancom. net
Centre Météorologique de Guadeloupe	Aéroport MARTINIQUE AIME CESAIRE	
CTC Aérogare Raizet sud - Zone Fret	Le Lamentin	
Rue Ernest Pallas	Service météorologique	
BP 18 - 97139 LES ABYMES CEDEX	97232 LE LAMENTIN CEDEX	

Figure M13.12

An extract from the French Aeronautical Information Publication (AIP) stating the official source of aviation meteorology information for France.

forecast weather is very often the biggest factor in deciding how (or if) a flight will proceed. In terms of Threat and Error Management (TEM), it is the weather that is always one of the 'Threat' factors to be considered. In many ways, weather epitomises a 'Threat' – it is an external factor which is outside the control of the pilot, yet which can decide whether a flight can be completed safely as planned, or if some alternative plan is required.

From all of the foregoing, it should be evident that checking the current and forecast weather before every flight is something that should become an instinctive part of every pilot's pre-flight routine, and the earlier you start to develop this habit, the better your understanding of meteorology will be.

Nevertheless, the legal phrase "*...all available meteorological information appropriate to the intended flight*" presents a challenge in itself. In the online era, the problem is likely to be not so much having trouble finding meteorological information, but rather filtering the torrent of data that is available online. Type 'London Heathrow TAF' into a popular online search engine, for example, and you are likely to get over 100,000 results. The question is, which of these will give accurate and up-to-date meteorological information?

The starting point for authoritative aviation meteorological should be the Aeronautical Information Publication (AIP) for the state you are flying in. Within the 'GEN' (General) section there will be a description of 'Meteorological Services'. This section will describe what types of aviation meteorological services are available in that state, any procedures which are specific to that state and the contact details for the organisation responsible for the providing aviation meteorological services in that state. In almost every European state, that organisation will have a free-to-access website (although registration is often required) from which authoritative aviation meteorological can be obtained and it pays to become familiar with this website.

Figure M13.13

An example of some of the weather information available on a popular flight planning app (in this case SkyDemon).

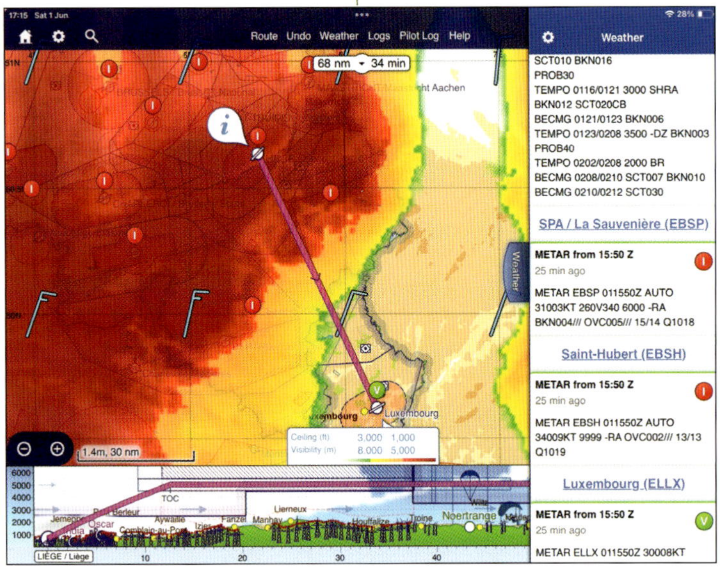

It is also worth knowing that TAFs and METARs in particular are freely shared between the various meteorological offices of the world, so it is not normally necessary to hunt-out each individual country's met. office in order to find the TAFs and METARs for that country's airfields – there will usually be a search function on your 'home' state's meteorological website allowing you to search for TAFs and METARs worldwide. However, it is much less likely that other aviation weather information (such as area forecasts and significant weather charts) will be widely – or freely – available outside the state originating that information.

In addition to designated met. offices, there are many websites that consolidate aviation weather information and make it available either for free, or via some form of subscription service. Increasingly there are also very powerful flight planning apps which will automatically display some relevant aviation weather information when the user plans a route.

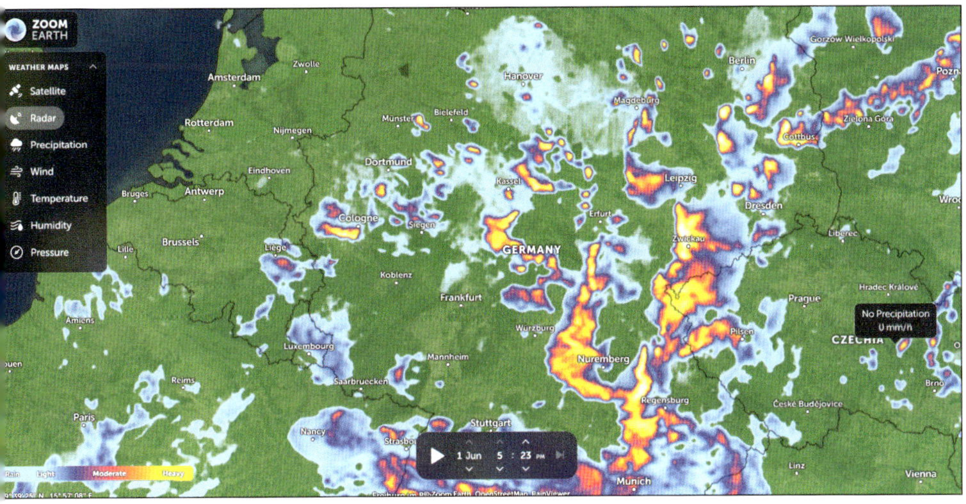

Figure M13.14
A website showing rainfall radar information.

Many weather websites may display more general weather information – rainfall radar, atellite images and surface pressure charts for example.

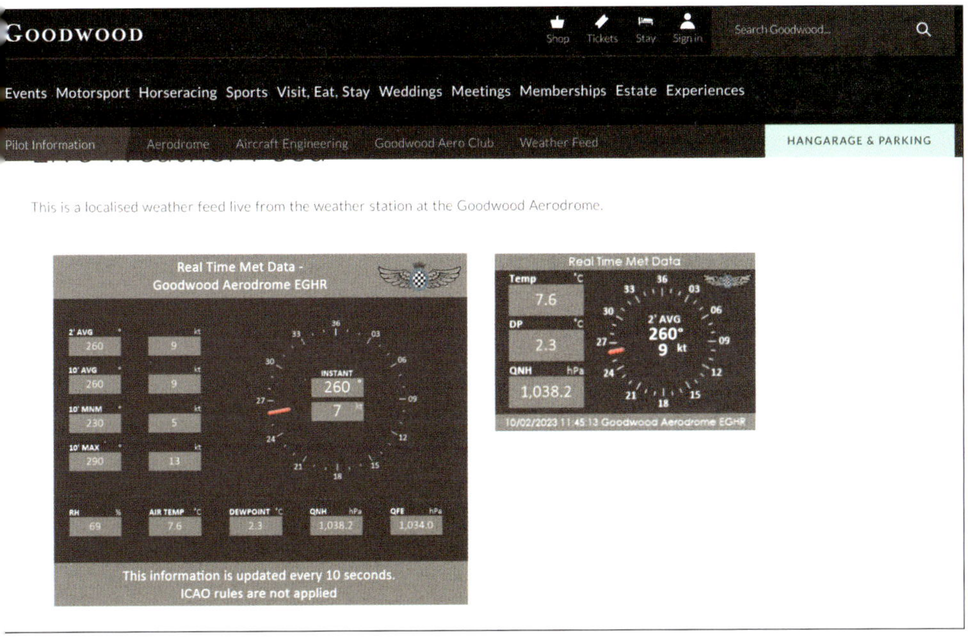

Figure M13.15
An airfield website with a weather observation.

More specifically, many airfields – especially smaller airfields with predominantly General Aviation users – may have some form of current weather information on their website. This current information is often obtained from the type of 'home weather station' that you can buy and install yourself at home if the mood takes you. Similarly, an airfield's website may display images from one or more webcams.

Away from the busier GA airfields, there are many amateur weather stations displaying

Figure M13.15a
An online image from an airfield webcam.

current weather information on-line, and there may be one of these near an airfield you wish to use. However, such weather information may be limited to basic parameters such as surface wind, pressure and average rainfall, and contain little or no data on items such as visibility, cloud base etc. which can be all-important to a pilot making a fly/no-fly decision.

The almost limitless aviation met. information available on-line does mean that in most instances it is possible to self-brief on the current and forecast conditions in great detail, drilling down to the sort of primary data that weather forecasters themselves might use if you really want to. However, the proliferation of weather information available on-line does not mean that the pilot can be completely indiscriminate about the data they use. Above all it is the pilot's responsibility to make sure that the information they use is not only appropriate for flight planning purposes, but also that it is up-to-date. If you are obtaining met. information from an officially designated aviation meteorological office, there is a reasonable expectation that it will be up-to-date. Nevertheless, you should as a matter of habit double-check the time and date of any forecasts and observations. If you are accessing data from a non-official source, it is almost certain that the infrastructure behind that website will be less robust and may not be monitored at all for accuracy and currency. If using such a website or service, it is even more important to check (and re-check) that the information you are accessing is as current as it can be. It may not be evident that the data being displayed is hours (or even days) out-of-date. Aviation met. professionals and their systems will reject any data for which they cannot establish the 'when and where' and pilots should follow their example; in short:

Never assume – check

Although the vast majority of aviation weather information is available and accessed on-line, the pilot may well find some weather information in 'hard copy' format. A flying school may print-off weather information at the start of the day and display it on a notice board for example; the school, or the airfield, or a 'handling agent', may well have a dedicated flight planning room or offer 'briefing packs' of pre-flight information. In all instances, however the information has been obtained, it is still the pilot's responsibility to establish the 'where and when' and validity of that information. Or, to repeat the point:

Never assume – check

An airfield may have an Automatic Terminal Information Service (ATIS) – a recorded radio broadcast of the observed weather and other aeronautical information for the airfield. The broadcast will start with the airfield name and the time of the broadcast, so straight-away the pilot can verify the 'where and when' of the information. Aside from accessing an ATIS broadcast by radio, at some airfields the ATIS broadcast can also be accessed via telephone, which can be particularly useful for checking the ATIS before travelling to the departure airfield, or for checking the ATIS at the destination airfield.

If flying to or from a small airfield or landing site, the only source of weather information may be a non-professional observer. Such an observer may offer an opinion on the visibility, based on the clarity (or otherwise) of local landmarks, or estimate the cloud base from the appearance of local topography. Such a person may even offer an opinion of how the conditions are likely to change – an unofficial forecast of sorts. These unofficial observations must be treated with a great deal of caution, even if (indeed especially if) they are telling the pilot something that they want to hear. An unofficial weather observation may come from a seasoned aviation professional with decades of experience weather-watching at that airfield. Or it may come from a lay observer with little or no relevant aviation experience.

There is the story of the pilot who rang the destination airfield and was told that the weather was lovely and calm, and that the sun was shining. On arrival over the airfield the ilot was able to see that the sun was indeed shining – through a thick layer of mist....

When considering weather during the pre-flight phase, there is a recommended

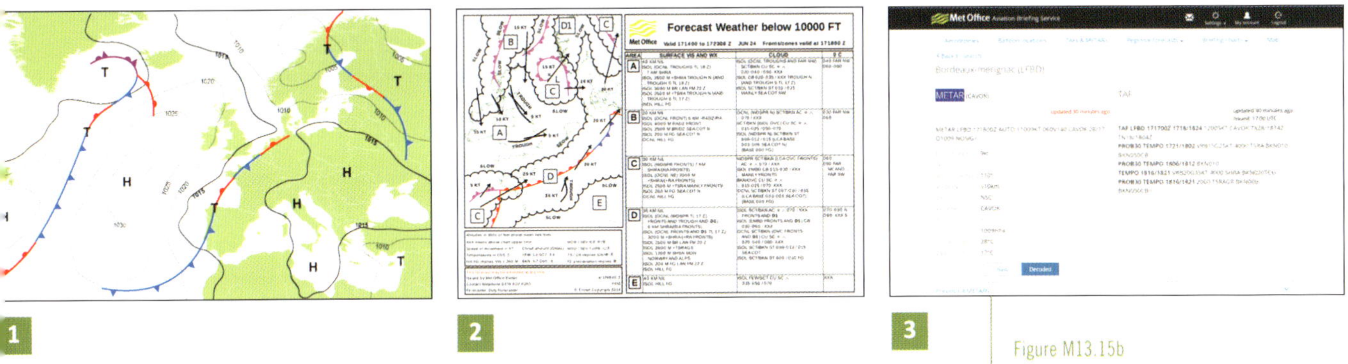

Figure M13.15b

When planning a flight, check the forecast and actual weather in the order:

1 General Weather Situation

2 Area Forecast

3 Airfield weather

equence of actions which should help the pilot to best understand the weather situation nd make safe and timely flight planning decisions.

Firstly, the general weather situation for the egion can be deduced from charts of pressure listribution showing highs (anticyclones), lows depressions), fronts, troughs etc. These charts re widely available not just on aviation weather vebsites but also on other media such as TV veather forecasts and reports. Sometimes called 'synoptic' weather charts, forecast charts or several days into the future are widely vailable and allow the pilot to see how the egional weather situation is developing. Knowing the general characteristics of ignificant weather features, the pilot can make ome broad assumptions about the flying conditions likely to occur.

Secondly, once the pilot has an overview of he regional weather situation, area forecasts

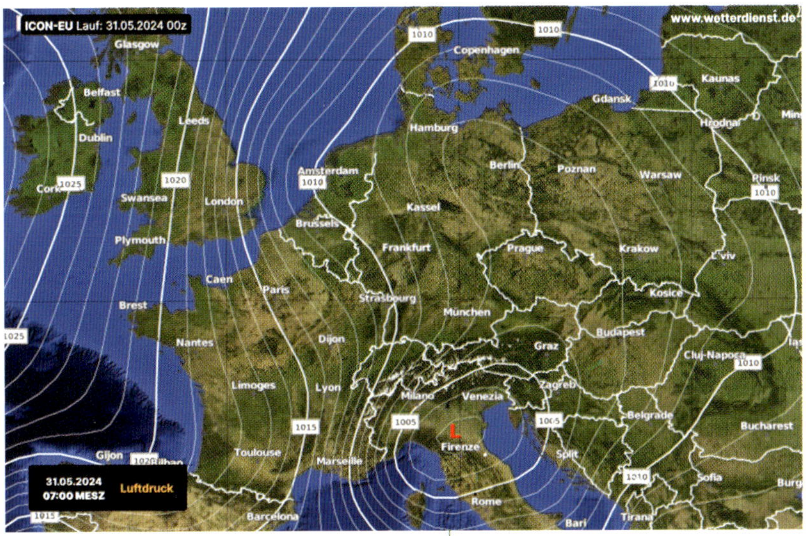

Figure M13.15c

A synoptic chart for Europe.

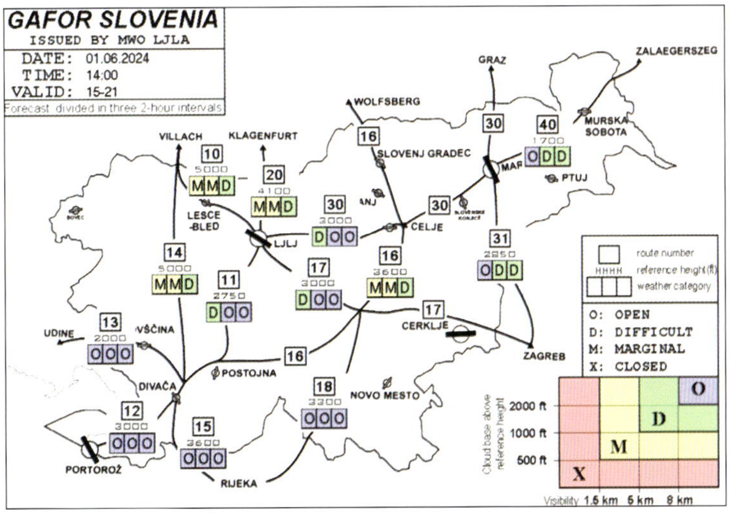

Figure M13.15d

An area forecast for General Aviation, in this case in Solvenia.

can be considered. Most states have some form of aviation-specific area weather forecast, covering some or all of the state's territory. These area forecasts may be text-based, chart-based, or a combination of the two. In any event, an aviation area forecast should use recognised aviation meteorology symbols and abbreviations and allows the pilot so see how the general weather situation portrayed by the regional pressure distribution charts is influencing the weather in the area of the planned flight. Other weather information – such as satellite images, rainfall radar etc. – can be used together with the area forecast to see how closely the actual weather is matching that forecast.

Finally, the pilot can consider whatever TAFs and METARs are available for the departure en-route and destination airfields, or those close to the departure or destination points. Other weather station reports, webcams and unofficial observations may also be used to assess weather at specific locations, subject to the limitations of this information already discussed.

Europe

EHAM **AMSTERDAM SCHIPHOL AIRPORT**
METAR EHAM 011625Z 01011KT 330V060 8000 SCT007 BKN008 15/14 Q1020 TEMPO 7000=
TAF EHAM 011109Z 0112/0218 36012KT 9999 SCT005 BKN008 BECMG 0115/0118 SCT008 BKN012 BECMG 0203/0206 SCT020=
EBAW **ANTWERPEN/DEURNE**
METAR EBAW 011620Z 33008KT 300V360 8000 OVC011 15/13 Q1020 TEMPO BKN008=
TAF EBAW 011420Z 0115/0124 33008KT 7000 SCT009 BKN014 PROB30 TEMPO 0115/0124 BKN009=
LFOB **BEAUVAIS-TILLE**
METAR LFOB 011630Z AUTO 35010KT 9999 OVC012 13/11 Q1022 TEMPO 3000 RADZ BKN007=
TAF LFOB 011100Z 0112/0212 34012KT 9999 BKN012 OVC016 TEMPO 0112/0208 BKN007 PROB40 TEMPO 0112/0118 3000 RADZ BECMG 0208/0211 SCT020=
ENBR **BERGEN-FLESLAND**
METAR ENBR 011620Z 32017KT 290V360 9999 FEW040 15/10 Q1022 NOSIG RMK WIND 1200FT 33028KT=
TAF ENBR 011100Z 0112/0212 33010KT 9999 FEW015 SCT025 TEMPO 0112/0121 33015G25KT=
EKBI **BILLUND**
METAR EKBI 011620Z 31008KT 280V340 9999 FEW030TCU BKN070 22/17 Q1016=
TAF EKBI 011106Z 0112/0212 02006KT 9999 FEW035 TEMPO 0112/0118 4000 SHRA SCT030CB PROB40 TEMPO 0113/0118 TSRA=
LFBD **BORDEAUX-MERIGNAC**
METAR LFBD 011630Z AUTO 01010KT 340V040 9999 FEW040 SCT092 21/11 Q1020 TEMPO 3000 SHRA BKN020TCU=
TAF LFBD 011100Z 0112/0218 02010KT 9999 BKN030 TX22/0114Z TN11/0205Z TEMPO 0112/0118 3000 SHRA BKN020TCU BECMG 0118/0120 CAVOK BECMG 0208/0210 BKN020 PROB40 TEMPO 0213/0216 SCT020TCU=

Figure M13.15e
A list of TAF (forecasts) and METAR (actual reports).

By checking the weather in this sequence (region, area of flight, specific locations) the pilot should be able to plan the flight armed with a sound knowledge of the general weather situation, as well as actual and forecast conditions at specific locations. Desirable attributes (such as situational awareness) and the consideration of the 'threat' element of Threat and Error Management (TEM) are greatly enhanced when the pilot understands both the 'big picture' of the day's weather as well as the detail of the weather occurring and forecast to occur at defined locations.

Once the flight has commenced, there are still plenty of options for checking the weather along the route, and at the destination and possible diversion airfields.

Most European states have one or more **VOLMET** broadcasts. A VOLMET broadcast is a collection of recorded METAR reports for a group of airfields, usually within a specific area whose approximate region is given by the VOLMET name, eg London South, Marseille etc. A VOLMET broadcast can usually be received over a wide area and, in the case of a VOLMET broadcast on a VHF frequency, the higher you fly the greater the range of the VOLMET broadcast. During a long cross-country flight VOLMET can be a very good way of keeping up-to-date with the weather conditions along the route or at or near the destination. Aeronautical Information Publications (AIP) and commercial flight guides will publish VOLMET frequencies and the list of airfields contained in that broadcast (the airfield reports should always be broadcast in the same order). It can take several minutes for a VOLMET broadcast to run through its full cycle and it is one of the truisms of flying that when you do tune into a VOLMET broadcast, the airfield you are interested in has just been broadcast and you will have to wait for the whole cycle to complete for it to come around again.

Most larger airfields have an **Automatic Terminal Information Service** (**ATIS**) broadcast.

GEN 3.5 - 10 AIP IRELAND
08 OCT 2020

Name of Station	CALLSIGN/ IDENT/ abbreviation (EM)	Frequency	Broadcast Period	Hours of Service	Aerodromes / Heliports included	REP, SIGMET INFO FCST & Remarks
1	2	3	4	5	6	7
DUBLIN	Dublin VOLMET	127.005 (8.33 kHz Channel)	Continuous	H24	Reports for: Dublin, Shannon, Cork, Belfast, Glasgow, Preswick, Manchester, London/Heathrow, London/Gatwick,	Plain Language (En) Broadcast continuous through the 24 hours. Reports, plus landing forecasts in the TREND form are broadcast in the sequence listed. The reports broadcast are the latest available. If a new report is not available the earlier one will be reported for one hour only, together with time of observation.

The above table indicates the current VOLMET Broadcast content and sequence

SIGMET denotes the time-blocks during which SIGMET'S are broadcast.

Figure M13.16

A Volmet service for Ireland, as listed in the Ireland AIP.

Aerodrome Terminal Information Service (ATIS)

Certain aerodromes provide weather and/or aerodrome information by telephone, may not be available as a 24 hour service:

Airfield	Frequency	Telephone No
Aberdeen	121.855	
Barkston Heath	132.125	01400 231138
Belfast Aldergrove	126.130	
Belfast City	124.580	02890 935124
Benbecula	113.950	01870 604818
Biggin Hill	135.680	
Birmingham	136.030	0121 767 1260
Blackpool	127.200	
Boscombe Down		01980 663101
Bournemouth	133.725	01202 364151/364158
Bristol	126.030	01275 475686
Brize Norton	126.500	01993 897142
Cambridge	134.605	01223 295631
Cardiff	132.480	01446 729319
Cranfield	121.880	
Cranwell	126.325	
Dundee	119.330	01382 662222
East Midlands	122.680	
Edinburgh	131.355	0131 333 6216
Exeter	119.325	01392 444523
Fairford	135.455	01285 714805 Ex 4554
Farnborough	128.405	09063 800400 Ex 282
Glasgow	129.575	0141 887 7449
Gloucestershire	127.480	
Guernsey	109.400	01481 238957
Hawarden	125.430	
Humberside	124.130	01652 682020
Inverness	109.200	01667 464255
Isle of Man	123.880	01624 821642
Jersey	134.680	01534 446301
Kirkwall	108.600	01856 878476
Leeds Bradford	118.030	
Leeming		01677 423041 Ex 7770
Leuchars		01334 857829
Liverpool	124.330	
London City	136.355	0207 511 6055
London Gatwick	136.525	
London Heathrow	121.935 128.080	
London Luton	120.580	01582 395225
London Stansted	127.180	01279 669325
Londonderry	119.380	028 718 0442
Lossiemouth		01343 817666
Lydd	129.230	01797 322422
Manchester	128.180 121.980	0161 209 2860
Marham		01760 337261 Ex 7888
Merryfield		01935 455456
Middle Wallop		01264 784142
Newcastle	118.380	0191 214 3400/3401
Newquay	127.405	01637 861319
Northolt	125.125	0208 845 2300 Ex 8937
Norwich	128.630	
Oban		01631 572910
Oxford	136.230	
Prestwick	121.130	
Redhill	125.305	01737 822947
St Athan	130.560	01446 243943
Shawbury		01939 250351 Ex 7574
Shoreham	130.980	0844 2254201
Southampton	130.880	02380 625877
Southend	136.055	
Stornoway	115.100	01851 707444
Sumburgh	125.855	01950 461037
Teesside	132.380	
Valley	120.725	
Waddington		01522 727305
Warton	121.730	01772 856060
Wick	113.600	01955 607596
Yeovilton		01935 455648

Figure M13.17

A page from a VFR flight guide listing ATIS available by radio and telephone.

s already described, an ATIS broadcast will give current weather onditions at the airfield – a spoken METAR – and in addition will nclude any relevant airfield information such as runway in use, nserviceability of facilities, initial contact frequency etc. Whenever he ATIS broadcast is refreshed, it is given a new letter, for example: *This is Manchester information Delta...*". When first making contact vith the airfield, it saves time to acknowledge that you have eceived the ATIS broadcast, so that the controller does not have to epeat the relevant information to you. At some airports, there may be separate arrival and eparture ATIS broadcasts. At an airfield where METARs are produced, but there is no ATIS roadcast, the Air Traffic Service Unit (ATSU) – if there is one – will normally have the most ecent METAR to hand and be able to relay it to the pilot.

Figure M13.18

The AIP entry for Ostend airport, listing the ATIS radio frequency.

BELGIUM & LUXEMBOURG Aeronautical Information Publication
See cover page for details

PDF Help Feedback

AIP AMDT SUPs AICs Search

Effective 16 MAY 2024

Part 1 GENERAL (GEN)
GEN 0 INTRODUCTION
GEN 1 NATIONAL REGULATIONS AND REQUIREME
GEN 2 TABLES AND CODES
GEN 3 SERVICES
GEN 3.1 Aeronautical Information Services
GEN 3.2 Aeronautical Charts
GEN 3.3 Air Traffic Services
GEN 3.4 Communication Services
GEN 3.5 Meteorological Services
GEN 3.6 Search and Rescue
GEN 4 CHARGES FOR AERODROMES/HELIPORT:
Part 2 EN-ROUTE (ENR)
ENR 0 INTRODUCTION
ENR 1 GENERAL RULES AND PROCEDURES
ENR 2 AIR TRAFFIC SERVICES AIRSPACE
ENR 3 ATS ROUTES
ENR 4 RADIO NAVIGATION AIDS / SYSTEMS
ENR 5 NAVIGATION WARNINGS
ENR 6 EN-ROUTE CHARTS
Part 3 AERODROMES (AD)
AD 0 INTRODUCTION
AD 1 AERODROMES/HELIPORTS - INTRODUCTIO
AD 2 PUBLIC AERODROMES
+ EBAW ANTWERPEN / Deurne
+ EBBR BRUSSELS / Brussels-National
+ EBCI CHARLEROI / Brussels South
+ EBKT KORTRIJK / Wevelgem
+ EBLG LIÈGE / Liège
+ ELLX LUXEMBOURG / Luxembourg
- EBOS OOSTENDE-BRUGGE / Oostende
AD 2.1 EBOS Aerodrome Location Indicator and I
AD 2.2 EBOS Aerodrome Geographical and Admi
AD 2.3 EBOS Operational Hours
AD 2.4 EBOS Handling Services and Facilities

Open application menu

7 Remarks UAS can be encountered in UAS geographical zones EBOS VLL0, VLL1 and VLL2 (for specifications, see ENR 5.1 § 4). Systematic tracking of UAS by ATC cannot be ensured

EBOS AD 2.18 ATS COMMUNICATION FACILITIES

Service designation	Call sign	Frequency/Channel	Hours of operation	Remarks
1	2	3	4	5
APP / TAR	Oostende Approach	120.600 MHZ	H24	Primary frequency
		266.075 MHZ	H24	NIL
		121.500 MHZ 243.000 MHZ	H24	Emergency frequency
		127.330	H24	Supplementary frequency 8.33 KHZ CH
TWR	Oostende Tower	118.180	H24	Primary frequency 8.33 KHZ CH
		266.075 MHZ	H24	NIL
		121.500 MHZ 243.000 MHZ	H24	Emergency frequency
		127.330	H24	Supplementary frequency 8.33 KHZ CH
	Oostende Ground (1)	121.980	H24	8.33 KHZ CH
		127.330	H24	Supplementary frequency 8.33 KHZ CH
ATIS	Oostende Information	126.130	H24	8.33 KHZ CH D-ATIS AVBL (see GEN 3.4, § 3.4.2)
VDF	Oostende Homer	120.600 MHZ 121.500 MHZ	H24	NIL
		118.180	H24	8.33 KHZ CH
		127.330	H24	Supplementary frequency 8.33 KHZ CH

(1) Ground movement control

Flight Planning

In theory, a **Flight Information Service** (**FIS**) covering a Flight Information Region should also be able to provide weather information to the pilot on request. This can be particularly useful if flying to a remote airfield where there are no other sources of in-flight weather information available. In practice, whilst a relatively quiet FIS frequency is a good bet for such a request, on busy FIS frequencies there is no guarantee that the FIS controller will have the spare capacity to obtain and broadcast this type of information (which may require the controller to telephone the airfield in question). The pilot should never be afraid to ask for information that will aid the safety of the flight, but must also be realistic about the controller's capacity to handle such requests.

Other pilots may be a useful source of in-flight weather information, especially hazards such as thunderstorms, for example. However, just as when contacting a small airfield for unofficial weather information, all pilot reports must be treated with a certain degree of caution. What constitutes 'good visibility' to an experienced professional pilot may be very different to what a student pilot considers to be 'good'!

Some aircraft avionics are also capable of receiving real-time weather information. However, the availability of such data depends very much on the aircraft equipment and the appropriate broadcasting facilities. As with all aviation technology, this is a capability which is developing at different paces in different states and with different avionics manufacturers.

Whatever the source of in-flight weather information, the basic principle of the 'when and where' applies – the pilot must be very clear about what airport (or area) the report or forecast covers, and the time of the observation or validity period of the forecast. The pilot should not accept any weather information that cannot be qualified with these two basic data points and should not hesitate to get clarification if necessary. In other words:

Never assume – check

Forecast and observation parameters and tolerances

It is often said that meteorology is an inexact science. This mantra, and many other less complimentary sentiments, are used by pilots whenever the weather chooses not to follow the forecaster's predictions, or when the actual weather conditions look rather different to the pilot than to the met. observer. To get the best out of meteorological reports and forecasts, it is helpful for the pilot to know the acceptable range of measurements and predictions and so appreciate the limitations of aviation weather information.

In practical terms, a weather observation – such as a METAR – should be considered as very accurate. It is, after all, a factual report of the conditions currently occurring. According to ICAO guidelines, the reported surface wind direction should be within 10° of actual, the speed within 10% of actual speed (once wind speed is above 10 knots). Temperature and dew point should be accurate to within 1°C, and pressure to within 0.5hPa.

Other elements of a METAR have a wider range of acceptable accuracy. Once visibility is more than 1500m, the measured visibility is considered accurate to +/- 20%; hence a reported visibility of 8km is considered accurate within a range of 6.4km to 9.6km. Cloud base (above 300ft) is considered accurate over a range of +/- 10%, hence a reported cloud base of 2000ft would be considered accurate over a range of 1800ft to 2200ft.

In addition to these accuracy standards, it has to be remembered that an aircraft in flight close to an airfield may encounter conditions quite different to those being reported from the surface at the airfield. In-flight visibility in particular may be different to that being observed at the surface, and may vary greatly depending on the direction of observation and the position of the sun, for example.

It will probably not be a great surprise to know that the acceptable accuracy parameters of an aviation weather forecast are somewhat wider than those for weather observations.

\ccording to ICAO recommendations, the surface wind direction should be accurate to vithin +/- 20% and wind speed to within 5 knots. However, these accuracy standards are only expected to be achieved 80% of the time. In other words, it is acceptable for the wind lirection and speed to be outside these forecast tolerances 20% of the time, or on 1 in 5 occasions. In the same way, forecast visibility (between 800m and 10km) has an accuracy arget of +/- 30%, 80% of the time. So if visibility is forecast as 7km, the actual visibility :ould be anywhere between 4.9km and 9.1km, on 4 out of 5 occasions during the forecast)eriod, and still be considered accurate. The occurrence, or non-occurrence, of forecast)recipitation is considered accurate if it follows the forecast 80% of the time. Cloud heights)etween 1000ft and 10000ft are considered accurate if the actual cloud height is +/-30%, luring at least 70% of the relevant forecast period.

If the above limits are treated as an aspiration, a target, for weather forecasters, it .hould be clear that even a TAF that is considered accurate within ICAO standards can :ontain significant variations between the forecast and actual conditions – the weather is ınder no obligation to follow the forecast's predictions! Of course, a METAR or TAF may be ımended (AMD), corrected (COR) or re-issued if there is a significant variation observed or ›xpected; which underlines the point that the pilot should always take care to use the most ıp-to-date weather information available.

There is one additional point worth re-emphasising about forecast accuracy, and that is :oncerning the clearance of early morning fog. It seems to many pilots that the forecast :learance of morning fog is often over-optimistic and the fog frequently seems to clear more .lowly than forecast, or even not clear at all (especially in the winter months). Maybe it is he case that when atmospheric conditions such as temperature and dew-point are very inely balanced, a very small change in wind speed or upper cloud cover, for example, can nake all the difference to the clearance of fog. In any event, in such situations it is unwise to ely entirely on a forecast fog clearance when planning a flight and at the very least, the)roficient pilot will carry plenty of reserve fuel and have alternate plans of action to hand vhen there is fog or mist in the forecast.

Figure M13.19
Any forecast time of fog clearance, or other significant weather event, should be treated with caution by the proficient pilot.

Progress Check

80. 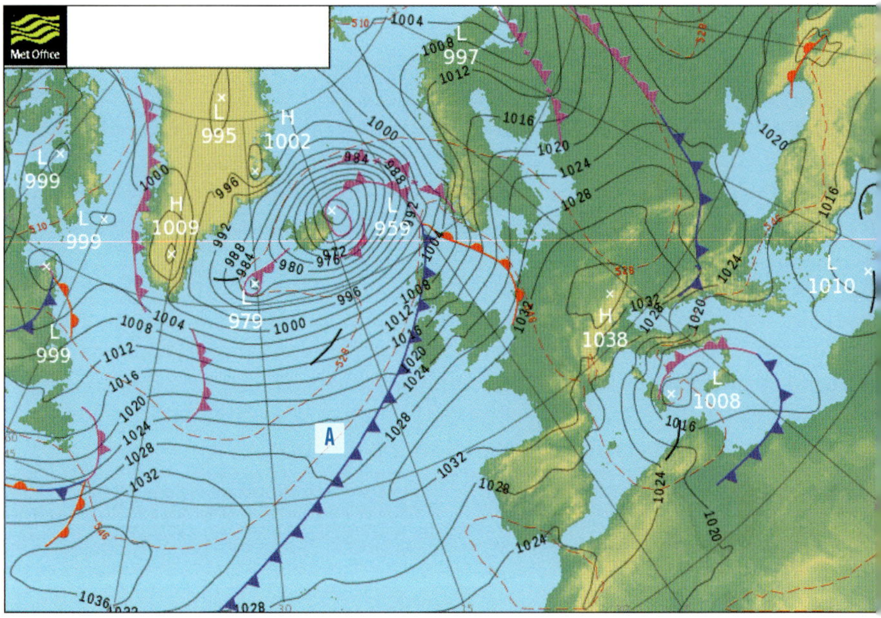

In the synoptic chart above, what weather feature is labelled 'A' and what is the sea level pressure at the centre of the depression it is associated with?

81. 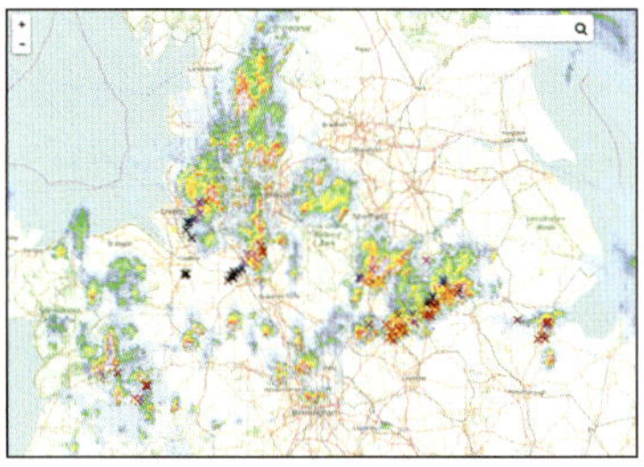

In the weather radar image above, what weather phenomena is indicated by the brightest yellow/orange/red colours?

82. On a significant weather (SIGWX) chart, what does the abbreviation EMBD stand for?
83. What is the difference between a METAR and a TAF?
84. Review the following METAR:
EGZZ 130420Z AUTO 03002KT 9999 FEW006/// SCT011/// BKN031/// 01/01 Q1003 REFZRA
What type of dangerous precipitation is being reported?
85. Review the following METAR:
EGZZ 191520Z 25043G58KT 3500 R27/0500 +RAGS TS SQ SCT010CB BKN013 06/06 Q0979 RETS
What surface wind and visibility is being reported?

86. Review the following TAF:
EGZZ 281038Z 2812/0106 10016G26KT 9999 FEW015 SCT030 PROB40 TEMPO 2812/2816 0600 +SHSN VV/// PROB40 TEMPO 2818/0106 2000 -SHSN SCT008
What weather conditions are being forecast to occur temporarily between 1200 and 1600 UTC
87. Review the following TAF:
EGXX 231637Z 2318/2418 12008KT 9999 FEW040 PROB40 TEMPO 2318/2322 4000 +TSRA TEMPO 2322/2402 3000 +SHRA TSRA PROB30 TEMPO 2322/2402 VRB15G25KT +TSGR BECMG 2400/2403 27012KT TEMPO 2402/2407 3000 BR BKN003 PROB30 TEMPO 2403/2406 0300 -DZ FG BKN001 PROB30 TEMPO 2407/2409 7000 BKN008 BECMG 2412/2415 21006KT
What condition are forecast to occur temporarily between 0300 and 0600 UTC?
88. Where can you check the official source of aviation meteorological information for a state?
89. How would a pilot normally obtain ATIS information?
90. What is the term used to describe a broadcast of recorded METAR reports for a group of airfields?

These questions are intended to test knowledge and reinforce some of the key learning points from this section. In answering these questions, a 'pass rate' of around 80% should be the target.

Model answers are found at page M177-M178.

Blank Page

M14 National Procedures

National Procedures

Progress Check

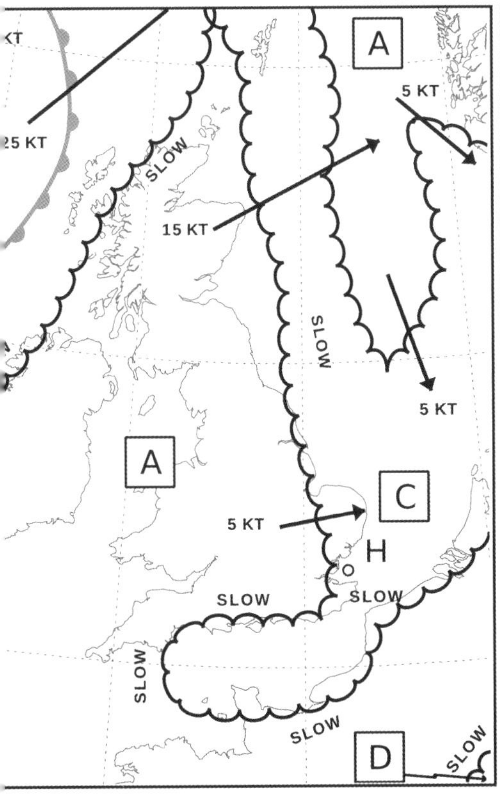

Forecast Weather below 10000 FT

Met Office Valid 080800 to 081700 Z MAY 24 Fronts/zones valid at 081200 Z

AREA	SURFACE VIS AND WX	CLOUD	0 C
A	35 KM NIL ISOL 7 KM SHRA LAN CONTINENT ... ... AND UK N FM 10 Z ISOL 3000 M BR LAN TL 09 Z ISOL HILL FG	OCNL SCT/BKN AC Ѱ ⋀ ... ... 070-090 / XXX FAR NW ISOL (AREAS LAN) SCT/BKN CU SC ... ... Ѱ ⋀ 015-030 / 035-050 (LCA ... ... XXX LAN CONTINENT AND ... ... UK N FM 10 Z) ISOL SCT/BKN ST 006-012 / 015 ... ... MAINLY SEA WINDWARD COT AND... ... UPSLOPES	020-040 ... ... FAR N ... ... AND ... ... FAR NE 060-080 NE XXX SW
B	20 KM NIL/-RA WDSPR 6 KM RA/-RADZ ISOL (OCNL FRONTS) 3000 M +RA/RADZ ISOL 2000 M DZ/BR WARM SECTOR OCNL ⋀ OCNL HILL FG	BKN/OVC AC AS Ѱ ⋀ 070 / XXX ISOL OVC NS Ѱ 025-035 / XXX FAR NW BKN/OVC CU SC (Ѱ FAR N) ⋀ ... ... 015-035 / 070 WDSPR (ISOL FAR NE) SCT/BKN ST ... ... 005-010 / 015 (LCA BASE 001-003 ... ... WARM SECTOR MAINLY NW)	040 FAR N 080
C	20 KM NIL OCNL 3000 M BR/-DZ ISOL 400 M FG SEA COT OCNL HILL FG	AREAS SCT/BKN SC ⋀ ... ... 015-030 / 040-060 OCNL SCT/BKN ST 005-012 / 015 ... ... (BASE 000 FG)	060-080

M14 National Procedures

As you might expect, the physical properties that make up meteorology are the same all over the globe because, as any fan of the original Star Trek TV series knows, you cannot change the laws of physics. Nevertheless, anyone who lives in, or spends time around, the British Isles knows that the weather of the UK is famously changeable and does from time-to-time defy even the best efforts of meteorological forecasters and their computer models. It follows that there are some features of the weather of the British Isles specifically which are worth noting.

M5 Clouds and Precipitation

Precipitation

In the UK, precipitation in the form of a mix of rain and snow, or 'wet' snow which turns to liquid on impact, is known as **sleet**, a term largely unknown outside the UK. In an aviation weather observation or forecast, such precipitation is described by the abbreviation RASN (**RA**in and **SN**ow).

M6 Visibility

Fog and Mist

Advection fog can form over the seas around the British Isles in winter months when warm air from the south is cooled below its dew-point by contact with the cold sea. An onshore breeze can carry such fog over land, and advection fog particularly affects the south-west UK and Ireland in the winter and spring when a south-westerly wind is blowing. Advection fog forming over the sea (sometimes called sea fog) can persist even in strong winds, unlike advection fog over the land which tends to be dispersed by strong winds. For example, in the Channel Islands thick fog combined with 40-knot winds is by no means rare! Advection fog can form over the land in winter when a relatively warm airflow (again usually from the south-west) passes over cold ground, perhaps with lying or thawing snow. The melting snow cools the air whilst adding moisture to it and widespread fog can form readily, often moving as a band at the leading edge of the warmer air.

Advection fog over the seas of the British Isles is worst during the spring and summer; over land it tends to be quickly dispersed by daytime heating and should not persist too far inland, although poor visibility and low cloud may be a problem just inland of the coast. That said, low cloud and advection fog can penetrate quite far inland overnight, particularly along river valleys. Thus airfields in valleys close to the coast (such as Edinburgh, Newcastle and Teesside in the UK) can be particularly vulnerable to low stratus and fog moving in from the sea, which reaches them earlier than other airfields and stays longer.

M7 Air Masses

Characteristics of Air Masses

Tropical maritime air is usually stable on reaching the British Isles. In winter, a tropical maritime air mass tends to bring dull weather with low stratus cloud, poor visibility and even advection fog (especially in south-west England). During the summer, surface heating effects cause cloud to dissipate inland and visibility to improve, although radiation fog becomes a possibility if the other necessary conditions prevail.

A tropical continental air mass is hot and dry and tends to be stable on reaching the British Isles. This type of air mass brings 'heatwave' weather to the British Isles in the summer with clear skies and high temperatures especially in Southern England. However, visibility is often poor due to haze and dust.

As polar continental air crosses the North Sea during winter, it can pick up enough moisture from the relatively warm sea to trigger off convective clouds and showers on eastern coasts of the UK. In the summer, cooling of polar continental air can result in banks of sea fog (known locally as "*Haar*" or "*sea fret*") over the North Sea. If conditions are right, this fog can move inland.

Arriving at the British Isles, polar maritime air brings convective (cumulus-type) clouds which are often well developed and produce showers or even thunderstorms. Outside these showers, visibility tends to be very good.

A 'returning polar maritime' air mass often brings low cloud to south-west regions of the British Isles in particular, but extensive, heavy showers and thunderstorms may move further across land areas, especially in the summer.

M13 Meteorological Information

Meteorological forecasting services for civil aviation in the United Kingdom are provided by the UK Met Office, whose public website (www.metoffice.gov.uk) contains many useful weather forecasting, reporting and educational resources.

An 'Aviation Briefing Service' is available free of charge for pilots flying within the UK (a simple registration process is required). Some extra services are available to those prepared to pay for the 'premium' service. Information available from the free site includes:

- Synoptic (surface pressure) charts
- Low Level Briefing Charts
- UK and near Europe TAF and METAR bulletins
- UK Ballooning Forecasts
- UK SIGMETs
- UK GAMETs
- London CTA helicopter forecasts
- UK Regional Pressure Settings
- UK observed weather map layers – visible and infrared satellite imagery and lightning, thunderstorms (day and night)
- UK high resolution rainfall radar
- (every 15 mins)
- Aerodrome Warnings email alerts

The UK Met Office issues synoptic charts forecasting up to three days ahead, and using the following symbols:

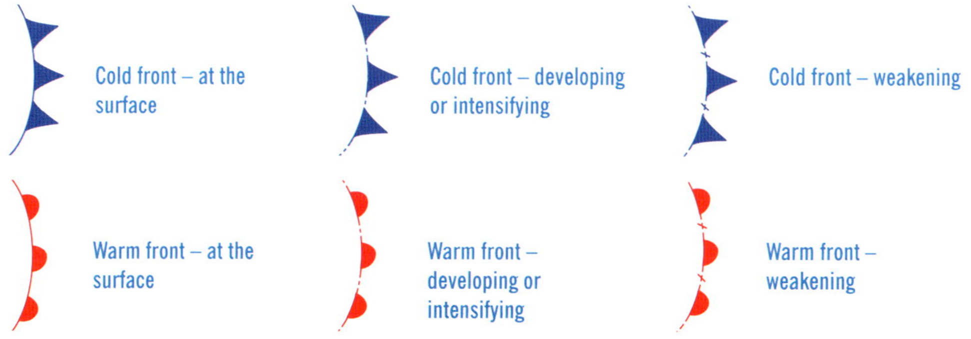

Figure M14.1
Symbols used on synoptic charts issued by the UK Met Office.

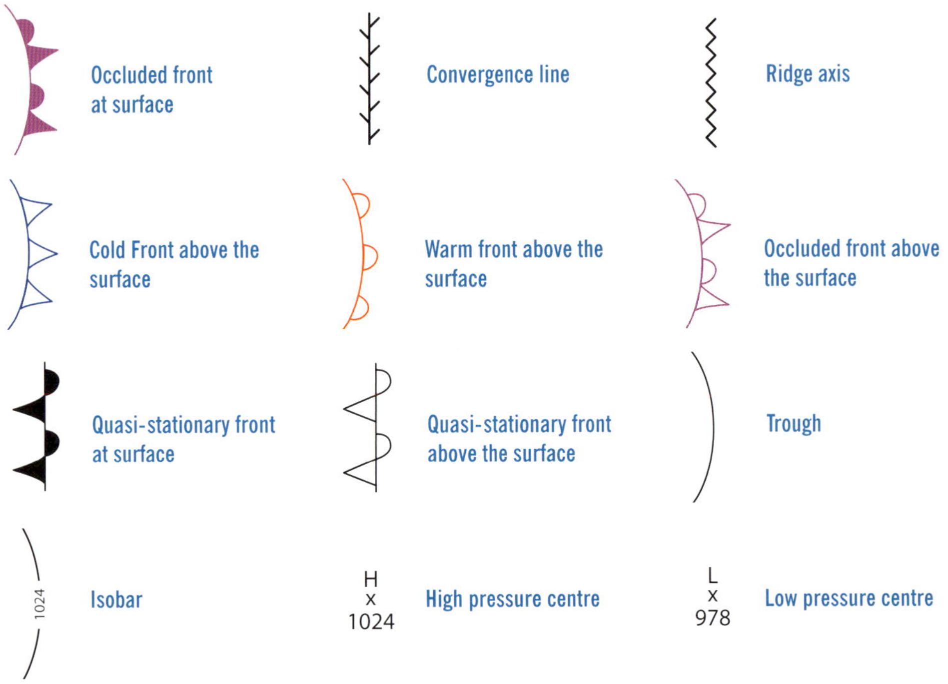

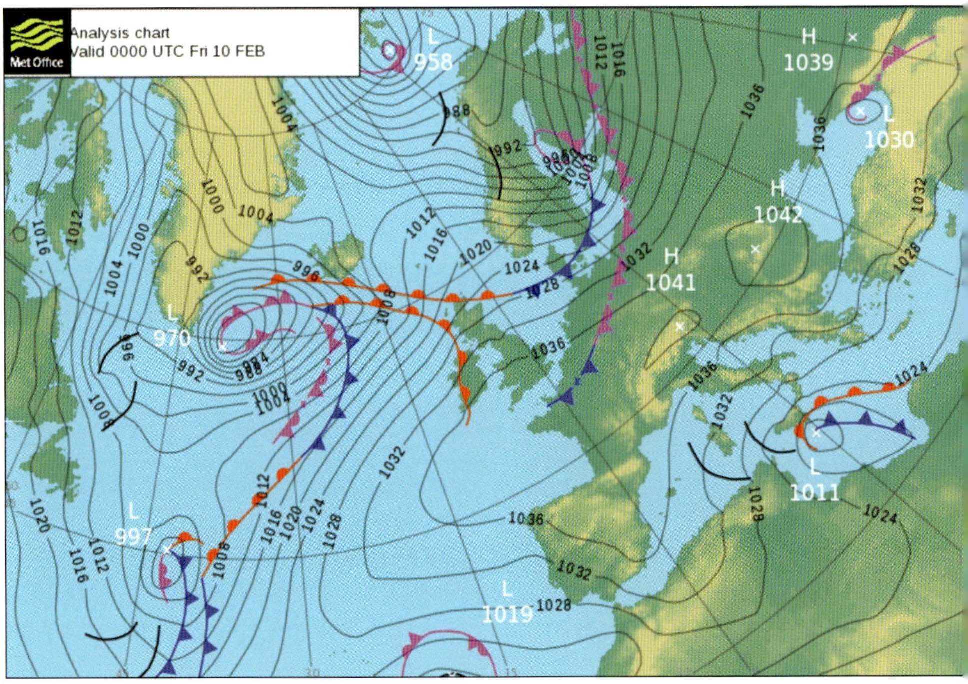

Figure M14.2
A Met Office synoptic chart, note the 'time and date' data top left.

The 'Aviation Briefing Service' section of the Met office website also gives free access to satellite images (visible and infrared), rainfall radar and lightning strike information.

Area and significant weather information available includes text SIGMETS; a specific forecast for helicopter operations in the London Control Zone / Control Area and forecast Regional Pressure Settings (regional QNH).

Met Office | **GAMET Area Forecast**

GAMET **South East Region**
Issued on Wednesday 08 May 2024 at 03:00 UTC

Valid for 0800-1700 UTC

Met Situation Valid at 1200 UTC
A SLOW-MOV AREA OF HIGH PRESSURE CENTRED NEAR MARGATE GIVES A LIGHT TO MOD N TO NE FLOW ACORSS THE REGION.

Strong wind warning
ISOL SEA COT MEAN 15KT.

Freezing Level
8000FT.

Weather Conditions

Zone 1

E OF A LINE KINGS LYNN TO SOUTHEND TO EASTBOURNE TO VENTNOR TO PLYMOUTH TO GUERNSEY TO LE HAVRE, THIS LINE GEN SLOW-MOV, BUT MOV E AT 5KT OVER EAST ANGLIA:
GEN 20KM, WITH 2-5/8ST 700FT/1500 AND 2-5/8SC 2000FT/4000.
OCNL 3000M IN BR, WITH 7-8/8ST 300FT/1500.
ISOL SEA COT, 400M IN FG, WITH 7-8/8ST SFC/1500.
WRNG: HILL FG. MOD TURB IN CLD.

Zone 2

REST OF REGION:
GEN 35KM, WITH 0-3/8ST 1200FT/1500 AND 2-6/8CUSC 3000FT/5000.
ISOL SEA WINDWARD COT AND UPSLOPES, 15KM, WITH 5/8ST 600-1200FT/1500 AND 0-4/8CUSC 2000FT/8000
ISOL LAN TL 09Z, 3000M IN BR, WITH NIL CLD.
WRNG: HILL FG. MOD TURB IN CLD.

Wind

Norwich			
Height (above ground level) / time (UTC)	0900	1200	1500
1000ft	010 5 KT +07	CALM +09	080 5 KT +11
3000ft	020 10 KT +08	020 10 KT +08	010 5 KT +08
6000ft	020 10 KT +03	020 10 KT +03	010 5 KT +03

Southampton			
Height (above ground level) / time (UTC)	0900	1200	1500
1000ft	130 5 KT +11	160 5 KT +13	190 5 KT +13
3000ft	010 5 KT +09	360 5 KT +09	360 5 KT +10
6000ft	340 5 KT +02	330 10 KT +03	340 10 KT +04

Oxford			
Height (above ground level) / time (UTC)	0900	1200	1500
1000ft	170 5 KT +11	200 5 KT +14	250 5 KT +16
3000ft	CALM +08	240 5 KT +08	260 5 KT +10
6000ft	310 10 KT +02	300 10 KT +02	280 10 KT +03

Gatwick			
Height (above ground level) / time (UTC)	0900	1200	1500
1000ft	030 5 KT +10	040 5 KT +13	030 5 KT +14
3000ft	030 5 KT +07	360 5 KT +07	350 5 KT +09
6000ft	340 10 KT +03	350 10 KT +03	350 10 KT +03

Regional Outlook (Valid 1800-2400 UTC)
ISOL BR AND FG LAN S FORMING FM 21Z. OTHERWISE, SIMILAR.

UK Extended Outlook (Valid 0000-2400 UTC)
HIGH PRESSURE WILL GIVE PREDOMINANTLY FINE WX. HOWEVER, EXTENSIVE ST AND ISOL BR/FG IS LIKELY TO AFFECT SOME COASTAL AREAS AT TIMES THROUGH THE PERIOD. AT 00Z, A FRONTAL ZONE LIES FROM THE NORTH SEA ACROSS N SCOTLAND TO N OF NORTHERN IRELAND, WITH OCNL RA OR DZ AND WDSPR ST. THIS FRONTAL ZONE CONTINUES TO MOV E MOV E THROUGH THE PERIOD.
WINDS GEN LIGHT AND VRB, BUT MOD TO STRONG SW WINDS IN THE FAR NW, AND FAR N, EASING TO LIGHT TO MOD FROM DAWN.

Figure M14.4
A sample GAMET.

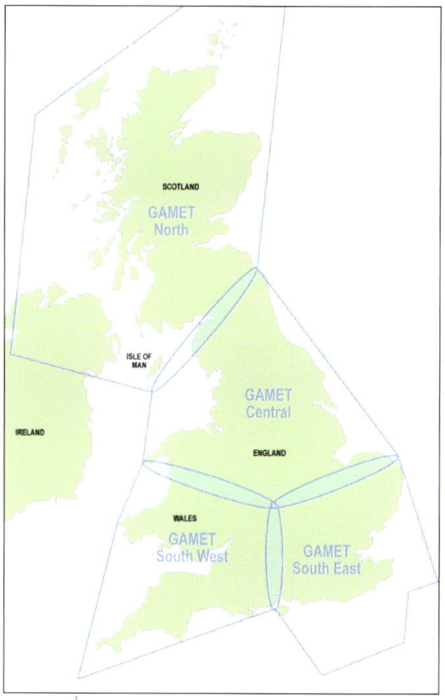

Figure M14.3
Met Office GAMET forecasts are available for four areas of the UK.

GAMET is a general aviation weather briefing service providing four area text forecasts overing the UK in plain language, which detail the current Met situation and also contain orecasts of wind, freezing level and weather conditions, and a forecast outlook. GAMET can e accessed via the Met Office website and uses standard meteorological abbreviations.

The UK Met Office also issues 'SIGMET' bulletins, which are text forecasts of significant in-flight weather phenomena, using standard meteorological abbreviations.

Figure M14.4

A sample SIGMET warning.

EGTT LONDON

Phenomena: Embedded thunderstorm(s) Valid from: Tue 25 Jun 2024 – 05:00 UTC Valid to: Tue 25 Jun 2024 – 09:00 UTC

EGTT SIGMET 02 VALID 250500/250900

EGRR- EGTT LONDON FIR

EMBD TS FCST WI N5130 E00200 – N5107 E00200 – N5100 E00128 – N5045 E00033 – N5130 W00054 – N5312 W00009 – N5245 E00303 – N5130 E00200

TOP FL360 MOV NNE 20KT WKN=

Possibly the most-used General Aviation area forecasts in the UK are the UK low-level forecast charts, the most popular of which forecasts in-flight weather over the British Isles below 10,000 (the **F215 forecast**). This forecast uses a mixture of a chart of the British Isles, with synoptic and aviation meteorology symbols and text using standard met. abbreviations

Figure M14.5

The key features of a Metform 215 Forecast.

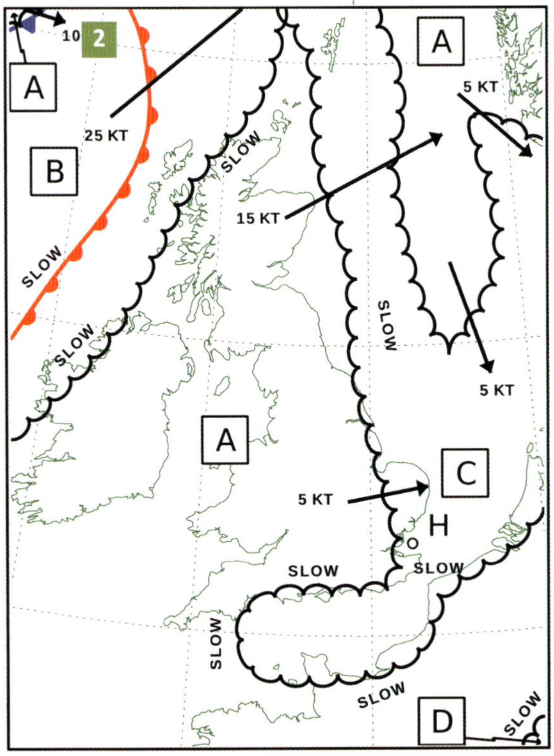

Altitudes in 100s of feet above mean sea level

XXX means above chart upper limit | MOD / SEV ICE

Speed of movement in KT | Cloud amount (Oktas) | MOD / SEV TURB

Temperatures in DEG C | FEW: 1-2 SCT: 3-4 | TS / CB implies GR/ /

Hill FG implies VIS < 200 M | BKN: 5-7 OVC: 8 | FZ precipitation implies

This forecast may be amended at any time.

Issued by Met Office Exeter at 080306 Z

Contact telephone 0370 900 0100 F215

Forecaster: Duty Forecaster

Met Office **1**

Forecast Weather below 10000 FT

Valid 080800 to 081700 Z MAY 24 Fronts/zones valid at 081200 Z

AREA	SURFACE VIS AND WX	CLOUD	0 C
A	35 KM NIL ISOL 7 KM SHRA LAN CONTINENT ... ... AND UK N FM 10 Z ISOL 3000 M BR LAN TL 09 Z ISOL HILL FG **3**	OCNL SCT/BKN AC ... ... 070-090 / XXX FAR NW ISOL (AREAS LAN) SCT/BKN CU SC ... ... 015-030 / 035-050 (LCA ... ... XXX LAN CONTINENT AND ... ... UK N FM 10 Z) ISOL SCT/BKN ST 006-012 / 015 ... ... MAINLY SEA WINDWARD COT AND... ... UPSLOPES **4**	020-040 ... ... FAR N ... ... AND ... ... FAR NE 060-080 NE XXX SW **5**
B	20 KM NIL/-RA WDSPR 6 KM RA/-RADZ ISOL (OCNL FRONTS) 3000 M +RA/RADZ ISOL 2000 M DZ/BR WARM SECTOR OCNL OCNL HILL FG	BKN/OVC AC AS 070 / XXX ISOL OVC NS 025-035 / XXX FAR NW BKN/OVC CU SC (FAR N) ... ... 015-035 / 070 WDSPR (ISOL FAR NE) SCT/BKN ST ... ... 005-010 / 015 (LCA BASE 001-003 ... ... WARM SECTOR MAINLY NW)	040 FAR N 080
C	20 KM NIL OCNL 3000 M BR/- DZ ISOL 400 M FG SEA COT OCNL HILL FG	AREAS SCT/BKN SC ... ... 015-030 / 040-060 OCNL SCT/BKN ST 005-012 / 015 ... ... (BASE 000 FG)	060-080
D	30 KM NIL ISOL 7 KM RA/SHRA ISOL 3000 M BR LAN TL 09Z OCNL HILL FG	ISOL SCT/BKN AC 080 / XXX OCNL SCT/BKN CU SC ... ... 015-035 / 050-080 ISOL SCT/BKN ST 008-012 / 015 ... ... LAN TL 09 Z	

Outlook Until 090000 Z: ISOL BR/FG LAN MAINLY FAR S FM 21Z, OTHERWISE SIMILAR. **6**

1 The validity time of the forecast in date and UTC. In this case 8th of the month, 0800 – 1700UTC. The position of weather features is valid at 1200 UTC on the 8th.

Forecast Weather below 10000 FT

Met Office   Valid 080800 to 081700 Z   MAY 24   Fronts/zones valid at 081200 Z

2 A map showing the major weather features, with direction and speed of movement. Scalloped lines enclose different areas, each is lettered and described in the forecast text.

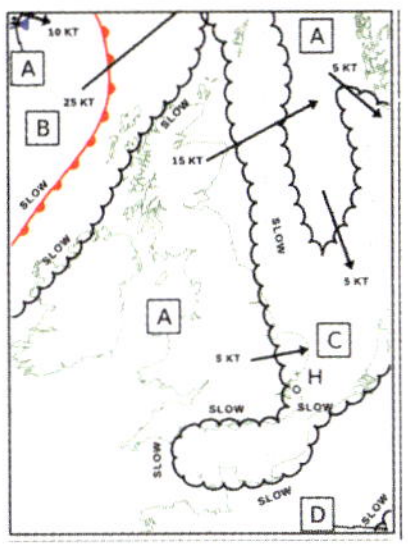

Each area of the map is given a letter, and sometimes sub-divided eg B1, B2 etc.

3 For each area the forecast surface visibility and weather is given using standard meteorology abbreviations.

4 For each area the forecast cloud amounts and levels are given; levels are in 100s of feet Above Mean Sea Level. 'XXX' indicates above forecast upper limit (10,000ft AMSL in this instance) Any significant icing and turbulence is also indicated.

5 For each area the forecast freezing level (0°C) is given.

6 The forecast outlook beyond the end of the chart validity period.

AREA	SURFACE VIS AND WX	CLOUD	0 C
A	35 KM NIL ISOL 7 KM SHRA LAN CONTINENT ... ... AND UK N FM 10 Z ISOL 3000 M BR LAN TL 09 Z ISOL HILL FG 3	OCNL SCT/BKN AC Ѱ ⋏ ... ... 070-090 / XXX FAR NW ISOL (AREAS LAN) SCT/BKN CU SC ... ... Ѱ ⋏ 015-030 / 035-050 (LCA ... ... XXX LAN CONTINENT AND ... ... UK N FM 10 Z) ISOL SCT/BKN ST 006-012 / 015 ... ... MAINLY SEA WINDWARD COT AND... ... UPSLOPES 4	020-040 ... ... FAR N ... ... AND ... ... FAR NE 060-080 NE XXX SW 5
B	20 KM NIL/-RA WDSPR 6 KM RA/-RADZ ISOL (OCNL FRONTS) 3000 M +RA/RADZ ISOL 2000 M DZ/BR WARM SECTOR OCNL ⋏ OCNL HILL FG	BKN/OVC AC AS Ѱ ⋏ 070 / XXX ISOL OVC NS Ѱ 025-035 / XXX FAR NW BKN/OVC CU SC (Ѱ FAR N) ⋏ ... ... 015-035 / 070 WDSPR (ISOL FAR NE) SCT/BKN ST ... ... 005-010 / 015 (LCA BASE 001-003 ... ... WARM SECTOR MAINLY NW)	040 FAR N 080
C	20 KM NIL OCNL 3000 M BR/-DZ ISOL 400 M FG SEA COT OCNL HILL FG	AREAS SCT/BKN SC ⋏ ... ... 015-030 / 040-060 OCNL SCT/BKN ST 005-012 / 015 ... ... (BASE 000 FG)	060-080
D	30 KM NIL ISOL 7 KM RA/SHRA ISOL 3000 M BR LAN TL 09Z OCNL HILL FG	ISOL SCT/BKN AC Ѱ ⋏ 080 / XXX OCNL SCT/BKN CU SC Ѱ ⋏ ... ... 015-035 / 050-080 ISOL SCT/BKN ST 008-012 / 015 ... ... LAN TL 09 Z	

Outlook Until 090000 Z: ISOL BR/FG LAN MAINLY FAR S FM 21Z, OTHERWISE SIMILAR. **6**

The Metform 215 provides a forecast for a nine hour period centred on a specified fixed time. It is designed to be used as either an area forecast or a route forecast.

When used in an F215 forecast, the following terms have specific meanings:

ISOL (isolated) – implies isolated conditions occurring randomly and which can easily be avoided, less than 25% of the area affected.

LCA (local or locally) – implies conditions clustered into small areas which can easily be avoided and is normally qualified geographically. Less than 25% of the area affected.

OCNL (occasional) – implies infrequent conditions which can be avoided, 25-50% of the area affected.

AREAS implies infrequent conditions clustered into areas which may be difficult to avoid. Places in between will be largely unaffected. 25-50% of the area affected.

FRQ (frequent) – implies conditions affecting many places which will be difficult to avoid. More than 50% of the area affected.

There is also a 'spot wind' chart (**F214**) forecasting wind velocities and temperatures at various levels over the British Isles.

Similar charts covering more of Western Europe are also available via the Met Office website.

The date and time period for which the forecast is valid.

For selected locations (given as a latitude and longitude) for forecast wind direction and speed, as well as the temperature, is given for different levels of as 1000s of feet; eg 02 = 2000ft, 10 = 10,000ft.

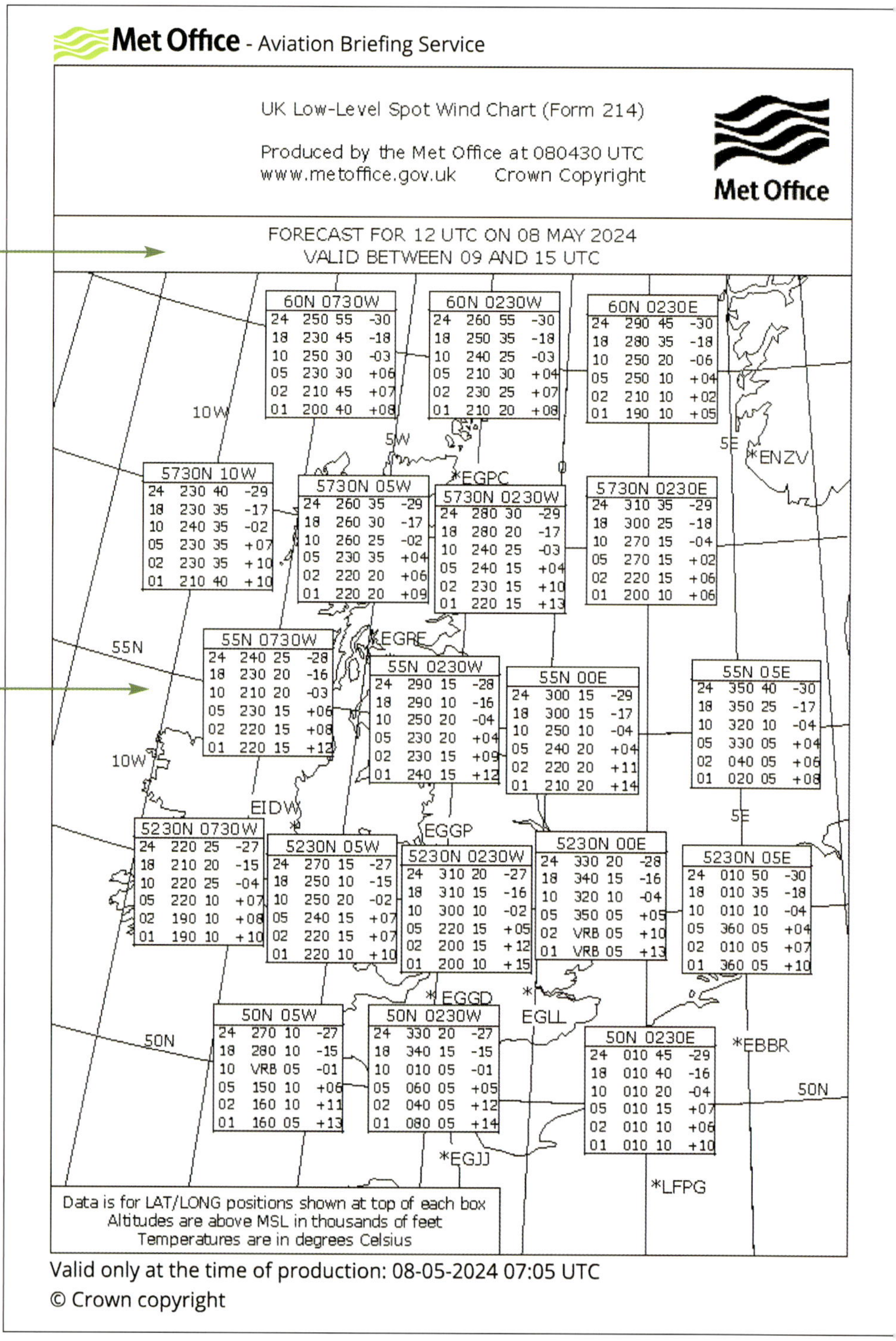

Figure M14.6

The key feature of a Metform 214 Forecast.

The UK Met Office also issues **Aerodrome Warnings**, which are made available to viation users, including airport operators, air traffic control units, and general aviation ilots. Warnings are displayed via the Aviation Briefing Service on the Met. Office website. Varning alerts by email are also available for registered users.

Lerwick/Tingwall Airport (EGET)
Warning issued!
Strong Wind
Issued at: Tue 07 May 2024 – 23:35 UTC Valid from: Wed 08 May 2024 – 10:00 UTC Valid to: Wed 08 May 2024 – 12:00 UTC
SSE WINDS WITH MEAN 20KT AT TIMES MAY OCCUR.

Figure M14.7

A sample Aerodrome Warning.

Aerodrome Warnings give advance notice of the following forecast conditions at specific irfields:

- Strong Wind (mean wind speed 20kt or more and/or gusts of 28kt).
- Gale (mean wind speed 34kt or more and/or gusts of 43kt).
- Frost (Ground Frost, Air Frost, Ground and Air Frost).
- Snow.
- Fog (including Freezing Fog).
- Thunderstorm (including Thunderstorm and Hail, Thunderstorm and Squall, Squall, Thunderstorm and Hail, and Squall).
- Hail.
- Freezing Precipitation (liquid precipitation freezing in contact with hard surfaces).
- Temperature Inversion (an increase in air temperature of 10°C or more in the lowest 1000ft).

Aerodrome Warnings will be issued if one or more of the specified phenomena occurs or is expected to occur. If there is a 40% or greater chance that one of the phenomena above will occur, the warning will say that the phenomena is *'likely to occur'*. If there is a less than 40% hance that the phenomena will occur, the warning will say that the phenomena *'may occur'*.

All Aerodrome Warning are provided in a standardised text format and warnings are ssued every six hours, four times per day. All warnings have at least a six-hour lead time, vith the exception of thunderstorm warnings that have a lead time of one hour. The varnings cover all or part of the six-hour periods. At a small number of airfields, warnings vill be issued as and when necessary.

In the UK, METAR observations are most commonly made at half-hourly intervals, at 20 minutes past the hour and 50 minutes past the hour (although some UK airfields, nostly military, METARs may be produced at hourly intervals). Surface wind neasurements contained in METAR reports are referenced to True North and are averaged over the previous 10 minutes, except when during the 10 minute period there vas an abrupt and sustained change in wind direction and/or speed. The measuring point used for the METAR wind information may be different to that used by Air Traffic Services at the airfield.

Where a UK METAR is generated by an automatic observing system without any human input or supervision and the present weather cannot be detected due to unserviceable or missing present weather sensors, the lack of present weather information will be indicated by two slashes (//). If the present weather sensor is unable to determine the state or form of the precipitation,

UP (unidentified precipitation) or **FZUP** (freezing unidentified precipitation), together with any intensity qualifiers, will be reported as appropriate. If the present weather sensor is serviceable but not detecting any present weather, then no present weather group is reported in the METAR.

Recent unidentified precipitation (**REUP**) is reported if moderate or heavy unidentified precipitation has ceased or decreased in intensity since being reported in the last METAR or within the last hour, whichever is the shorter.

In a METAR the term 'SNOCLO' indicates that the aerodrome is closed due to snow contamination of runways.

Military Aerodrome Colour State Code

Minimum weather conditions

Aerodrome Colour State	Surface visibility	Base of lowest cloud layer of 3/8 (or SCT) or more in heights above ground level (AGL)
Blue (**BLU**)	8km	2500ft
White (**WHT**)	5000m	1500ft
Green (**GRN**)	3700m	700ft
Yellow 1 (**YLO1**)	2500m	500ft
Yellow 2 (**YLO2**)	1600m	300ft
Amber (**AMB**)	800m	200ft
Red (**RED**)	Less than 800m	Below 200ft or sky obscured
Black (**BLACK**)	Runway unusable	

Figure M14.8
UK military airfield colour state codes.

UK military airfield METARs will usually report a **colour state** – a coded summary of the cloud base and visibility. The decode for these colour states is given in the table, but as the colour state code is additional to the normal METAR information, civilian pilots do not need to know these codings other than for their own information.

EGXT WITTERING

METAR EGXT 080820Z 17003KT 4000 BR OVC002 09/09 Q1029 RMK AMB=

Figure M14.9
A METAR for a UK military airfield; the colour code is Amber (AMB).

Military airfield weather colour codes are different to the airfield colour coding the UK Met Office uses on its weather map accessed via its website (although the cloud base / visibility criteria are relatively close for both systems). Some flight planning apps may also have their own colour coding system for weather information. In any event, checking the actual detail of the METAR will tell the pilot what the conditions are and how they relate to the pilot's personal limits.

A number of UK airfields have an ATIS service which can be accessed via a 'landline' telephone number.

In the UK, a TAF will not normally be issued until at least two consecutive METARs have been received from that airfield and accepted by the forecaster. Once issued, the TAF will be monitored against the METARs and amended if necessary. A UK TAF will be cancelled if:

– More than 3 half hourly METARs are not received;
– More than 3 half hourly METARs contain missing or erroneous data;
– A second hourly METAR is not received.

A TAF for a UK airfield will normally have a forecast period of 9 hours, 24 hours or 30 ours, but may be issued with a shorter duration if the airfield is due to stop issuing METARs efore the TAF duration would normally finish. A UK TAF is normally issued around 1 hour efore the start of its forecast period (for example, a TAF valid from 0600UTC to 1500UTC ill normally be issued around 0500UTC).

In the UK, 9 hour TAFs are re-issued every three hours; 24 and 30 hour TAFs are re-issued very six hours.

The UK Met Office can also supply a number of specialised aviation forecasts (for xample, for ballooning) and other specific forecast products, some of which are hargeable. Full details are available via the UK AIP (GEN section) and the UK Met Office ebsite.

Special aircraft observations must be issued by pilots as 'special air-reports' whenever ny of the following conditions are encountered or observed in UK airspace:

a) moderate icing (**MOD ICE**) or severe icing (**SEV ICE**);

b) moderate turbulence (**MOD TURB**) or severe turbulence (**SEV TURB**);

c) severe mountain wave (**SEV MTW**);

d) thunderstorms with or without hail (that are obscured, embedded, widespread or in squall lines) (**TSGR** or **TS**);

e) if volcanic ash cloud is observed or encountered.

pecial aircraft observations may also be reported when other meteorological conditions re encountered which, in the opinion of the pilot-in-command, might affect the safety of ircraft operations; for example adverse conditions during the climb-out or approach not reviously forecast or reported to the pilot-in-command.

The criteria for reporting en-route turbulence are based on the reaction of an airliner-ize aircraft.

urbulence Criteria

Incidence:

Occasional — less than 1/3 of the time

Intermittent — 1/3 to 2/3 of the time

Continuous — more than 2/3 of the time

ntensity	**Aircraft reaction (transport size aircraft)**	**Reaction inside aircraft**
ight	Turbulence that momentarily causes slight, erratic changes in altitude and/or attitude (pitch, roll, yaw). IAS fluctuates 5 – 15kt. Report as "*Light Turbulence*". or; turbulence that causes slight, rapid and somewhat rhythmic bumpiness without appreciable changes in altitude or attitude. No IAS fluctuations. Report as "*Light Chop*" Occupants may feel a slight strain against seat belts or shoulder straps. Unsecured objects may be displaced slightly. Food service may be conducted and little or no difficulty is encountered in walking.	
Moderate	Turbulence that is similar to Light Turbulence but of greater intensity. Changes in altitude and/or attitude occur but the aircraft remains in positive control at all times. IAS fluctuates 15 – 25kt. Report as "*Moderate Turbulence*". or; turbulence that is similar to Light Chop but of greater intensity. It causes rapid bumps or jolts without appreciable changes in altitude or attitude. IAS may fluctuate slightly. Report as "*Moderate Chop*". Occupants feel definite strains against seat belts or shoulder straps. Unsecured objects are dislodged. Food service and walking are difficult.	

Severe	Turbulence that causes large, abrupt changes in altitude and/or attitude. Aircraft may be momentarily out of control. IAS fluctuates more than 25kt. Report as "*Severe Turbulence*" Occupants are forced violently against seat belts or shoulder straps. Unsecured objects are tossed about. Food service and walking impossible.
	Note: The UK does not use the term 'Extreme' in relation to turbulence

Airframe Icing Intensity Criteria

Intensity	**Ice Accumulation**
Trace	Ice becomes perceptible. Rate of accumulation slightly greater than rate of sublimation. It is not hazardous even though de-icing/anti-icing equipment is not utilised, unless encountered for more than one hour.
Light	The rate of accumulation might create a problem if flight in this environment exceeds one hour. Occasional use of de-icing/anti-icing equipment removes/prevents accumulation. It does not present a problem if de-icing/anti-icing equipment is used.
Moderate	The rate of accumulation is such that even short encounters become potentially hazardous and use of de-icing/anti-icing equipment, or diversion, is necessary.
Severe	The rate of accumulation is such that de-icing/anti-icing equipment fails to reduce or control the hazard. Immediate diversion is necessary.
	Note: 'Rime Ice' is rough, milky, opaque ice formed by the instantaneous freezing of small supercooled water droplets. 'Clear Ice' is a glossy, clear, or translucent ice formed by the relatively slow freezing of large supercooled water droplets

Volmet

VOLMET	LONDON (MAIN)	LONDON (SOUTH)	LONDON (NORTH)	SCOTTISH
STATION	135.375	128.600	126.600	125.725
Aberdeen	■	■		
Amsterdam	■			
BELFAST-Aldergrove				■
Birmingham		■		
Bournemouth		■		
Bristol		■		
Brussels	■			
Cardiff		■		
Dublin	■			
East Midlands			■	
Edinburgh				■
Exeter		■		
Glasgow	■	■		
Humberside			■	
Inverness				■
Isle of Man			■	
Jersey		■		
Leeds Bradford			■	
Liverpool			■	
LONDON-Gatwick	■	■		
LONDON-Heathrow	■	■		
LONDON-Stansted	■			
London Luton		■		
Manchester	■	■		
Newcastle			■	
Norwich		■		
PARIS-Charles de Gaulle	■			
Prestwick				■
Southampton		■		
Southend		■		
Stornoway				■
Sumburgh				■
Teesside			■	

Figure M14.10
There are four VOLMET broadcasts in the UK.

There are four **VOLMET broadcasts** available in UK airspace: London Volmet Main, ondon Volmet South, London Volmet North and Scottish Volmet. The frequencies for hese broadcasts, and the airfields covered in each broadcast, can be found in the UK AIP nd commercial flight guides.

In UK VOLMET broadcasts, the spoken word '**SNOCLO**' will be added to the end of an irfield report when that airfield is unusable for take-offs and landings due to heavy snow n the runway(s), or runway snow clearance.

Surface wind sensors on UK airfields are positioned to give the best practical indication f the winds which an aircraft will encounter during take-off and landing within the layer etween 6 and 10m above the runway(s). The **surface wind reported** for take-off and anding by ATS Units at aerodromes supporting operations by aircraft whose maximum total veight authorised (MTWA) is below 5700kg is usually an instantaneous wind measurement vith direction referenced to Magnetic North. However, at other designated aerodromes the vind reports for take-off and landing are averaged over the previous 2 minutes. Variations n the wind direction are given when the total variation is 60° or more and the mean speed bove 3kt, the directional variations are expressed as the two extreme directions between vhich the wind has varied in the past 10 minutes. In reports for take-off, surface winds of kt or less include a range of wind directions whenever possible if the total variation is 60° r more. Variations from the mean wind speed (gust and lulls) during the past 10 minutes re only reported when the variation from the mean speed has exceeded 10kt. Such ariations are expressed as the maximum and minimum speeds attained.

Progress Check

91. What is the designated source of official aviation meteorological information in the UK?

92. 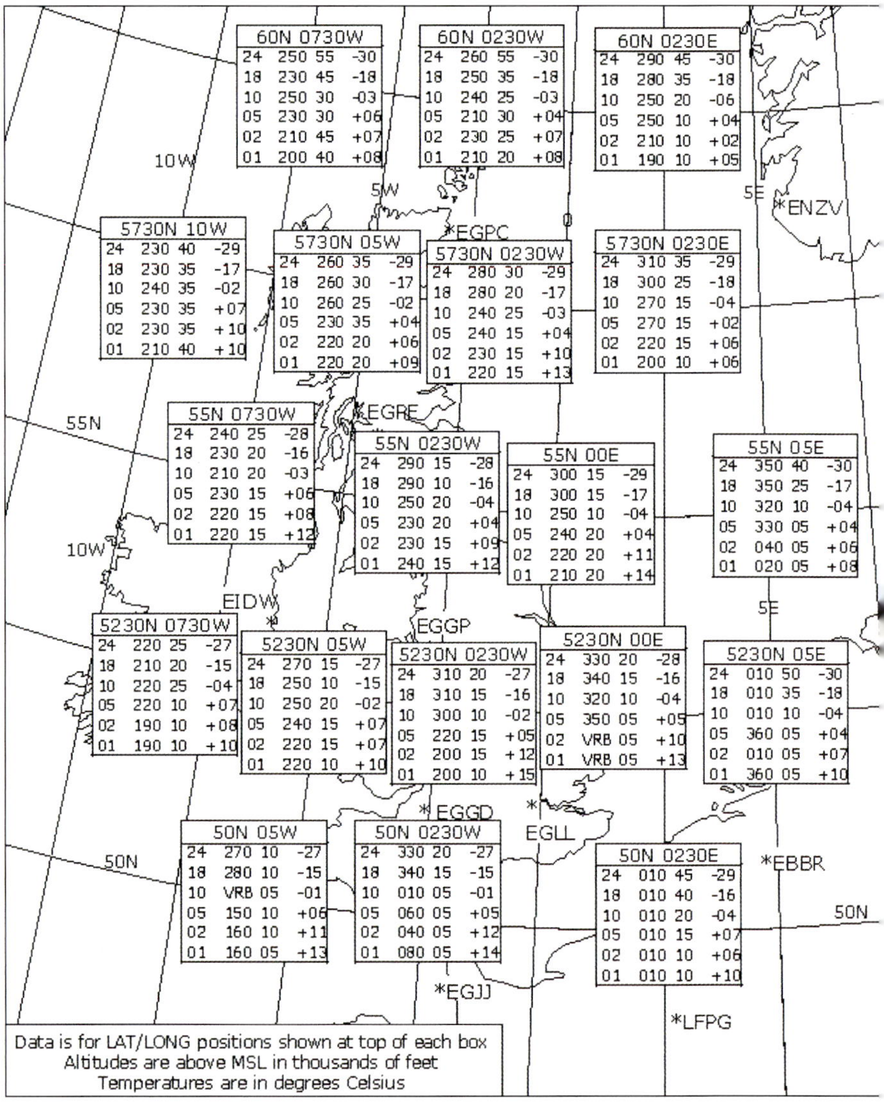

Refer to the Metform 214 chart. What is the forecast wind and temperature at 5000f at location 50N 0230W?

93. Refer the to Metform 215 chart below:

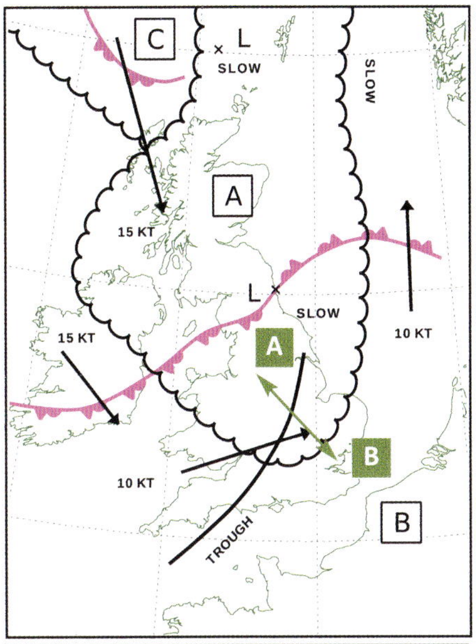

Altitudes in 100s of feet above mean sea level
XXX means above chart upper limit
Speed of movement in KT
Temperatures in DEG C
Hill FG implies VIS < 200 M
Cloud amount (Oktas)
FEW: 1-2 SCT: 3-4
BKN: 5-7 OVC: 8
MOD / SEV ICE Ѱ/Ѱ
MOD / SEV TURB ʌ/ʌ
TS / CB implies GR/Ѱ/ʌ
FZ precipitation implies Ѱ

This forecast may be amended at any time.
Issued by Met Office Exeter at 300914 Z
Contact telephone 0370 900 0100 F215
Forecaster: Duty Forecaster

Met Office

Forecast Weather below 10000 FT

Valid 301400 to 302300 Z MAY 22 Fronts/zones valid at 301800 Z

AREA	SURFACE VIS AND WX	CLOUD	0 C
A	25 KM NIL OCNL (FRQ FRONT AND TROUGH) ... ... 6 KM SHRA/(RA FRONT) ISOL 4000 M +SHRA/(+RA FRONT) ISOL 3000 M +TSRA ISOL 2500 M BR LAN FM 20 Z ISOL 200 M FG LAN FM 22 Z ISOL HILL FG	WDSPR SCT/BKN (LCA OVC ... ... TROUGH AND FRONT) AC ... ... Ѱ ʌ 080 / XXX WDSPR BKN/OVC CU SC Ѱ ʌ ... ... 015-035 / 080-XXX ISOL EMBD CB 015-030 / XXX ISOL SCT/BKN ST ... ... 007-012 / 015 (BASE 000 FG)	040-060
B	30 KM NIL ISOL (OCNL FRONTS AND TROUGH ... ... (AND LAN ELSEWHERE TL 19 Z)) ... ... 7 KM SHRA ISOL 4000 M +SHRA FRONTS AND ... ... TROUGH ISOL 2500 M BR LAN FM 20 Z ISOL 800 M SHSN MON NORWAY ISOL 200 M FG LAN FM 22 Z ISOL HILL FG	ISOL (WDSPR FRONTS AND TROUGH) ... ... SCT/BKN AC Ѱ ʌ 080 / XXX WDSPR (ISOL CONTINENT FM 19 Z) ... ... SCT/BKN CU SC Ѱ ʌ ... ... 020-040 / 070-XXX ISOL SCT/BKN ST 010 / 015 ... ... SEA COT N ISOL BKN ST 000 / 005 LAN FM 22 Z	050-070
C	25 KM NIL ISOL (WDSPR FRONT) 6 KM RA/SHRA ISOL 3000 M BR/RADZ OCNL HILL FG	WDSPR SCT/BKN (OCNL OVC FRONT) ... ... AC AS Ѱ ʌ 080 / XXX BKN/OVC CU SC Ѱ ʌ ... ... 015-025 / 060-080 WDSPR SCT/BKN 003-008 / 015	035-040

Outlook Until 310600 Z: SIMILAR.

A pilot is planning to fly from Manchester ('A') to Southend ('B'), departing from Manchester at 1600 UTC on the 31st May, with an expected flight time of 1:45. Assuming the pilot plans to maintain VMC, flying no lower than 1500ft AMSL while en-route, can the flight take place?

These questions are intended to test knowledge and reinforce some of the key learning points from this section. In answering these questions, a 'pass rate' of around 80% should be the target.

Model answers are found at page M178.

Blank Page

M Progress Check Answers

M1 The Atmosphere

M2 Temperature, Pressure and Density

M3 Altimetry

M4 Wind

M5 Clouds and Precipitation

M6 Visibility

M7 Air Masses

M8 Low Pressure Systems

M9 High Pressure Systems

M10 Hazardous Weather Conditions: Icing

M11 Hazardous Weather Conditions: Thunderstorms

M12 Other Hazardous Weather Conditions

M13 Meteorological Information

M14 National Procedures

M

Progress Check Answers

M1 The Atmosphere

1. In the theoretical 'dry' atmosphere, there is an average of 21% oxygen
2. Water vapour is the gas is the atmosphere which is responsible for almost all 'weather' -clouds, rain, snow etc.
3. The lowest layer of the atmosphere, extending from the surface to around 20,000ft over the poles and around 60,000ft over the equator, is known as the troposphere.
4. The layer at the top of the troposphere, separating it from the stratosphere, is the tropopause. Although the altitude of the tropopause varies from poles to equator, and varies from day to day, and varies from season to season, for practical purposes a General Aviation (GA) pilot can accept an average figure of 35,000ft (10,000m) over Europe. The actual level of the tropopause is given on some meteorological charts.

M2 Temperature, Pressure and Density

5. On average, air temperature reduces as altitude increases.
6. A temperature inversion is a situation where instead of air temperature reducing as altitude increases, it actually increases with increasing altitude.
7. As altitude increases, there is less weight of the atmosphere pressing down from above. Therefore, as altitude increases, pressure reduces. In the lower levels of the atmosphere near the surface, the rate of change of pressure is around 30ft per hectopascal.
8. A large area of low pressure is known as a depression.
9. The standard unit of measurement for air pressure in aviation is the hectopascal (hPa).
10. As altitude increases, air density reduces.
11. Density is inversely proportional to air temperature. Therefore, assuming pressure is constant, as temperature increases, density reduces.
12. As air temperature reduces, it can hold less water vapour (which is an invisible gas). Eventually, if temperature decreases sufficiently, the air becomes 'saturated' – it is holding all the water vapour it can.
13. The temperature at which air will become saturated is known as the 'dew point' temperature. The dewpoint varies from day to day and is reported in aviation meteorological observations. If air is cooled below its dewpoint, visible moisture (such as cloud or fog) will form.
14. In the International Standard Atmosphere (ISA), the surface air pressure is 1013.25 hectopascals (hPa).
15. In the International Standard Atmosphere (ISA), the surface temperature is 15°C and the lapse rate (the rate at which temperature reduces with increasing altitude) is 2°C per 1000ft. It follows that at 5000ft, temperature reduces by (5 x 2 = 10°C) in the ISA. 15°C – 10°C = 5°C.

M3 Altimetry

16. The altimeter is a form of barometer with a set of metal 'bellows' that have been partially evacuated and sealed. As air pressure rises and falls the bellows expand and contract. This movement is measured, calibrated, and displayed on the altimeter face in terms of vertical distance.

17. Altitude (vertical distance Above Mean Sea Level – AMSL) is displayed on an altimeter when it is set to QNH.

18. When a QFE pressure setting is used on an altimeter, the altimeter will display the height (vertical distance above a specific point on the surface) above a specific point – usually an airfield.

19. Flying from high pressure to low pressure, a specific pressure level will become closer to the surface. It follows that if an aircraft is following a pressure level (as will happen if it is maintaining the same indicated vertical distance without adjusting the altimeter pressure setting), it will become closer to the surface – *even though the indicated level on the altimeter is unchanged*.

    In this situation the altimeter is 'over-reading', it is reading a higher level than is correct. This situation can be remembered as:

    High to low: down you go

20. Assuming 30ft = 1hectopascal (hPa), a vertical distance of 300ft = 10 hPa. QFE is the pressure setting to indicate height above a specific point on the surface – normally an airfield. So, with 1027 (the QFE) set on the altimeter, the altimeter reads zero on the airfield.

    QNH is used to indicate altitude above sea level and assuming that an airfield is above sea level, QNH will be higher than QFE. 1027 + 10 = 1037hPa.

21. In a column of cold air, any given pressure level will be closer to the surface than in a column of warm air. So, if an aircraft follows a constant pressure level (hence a constant altimeter indication) from high temperature to low temperature, the aircraft will actually get closer to the surface: the altimeter is over-reading:

    High to low: down you go

M4 Wind

22. A line joining places of equal sea level pressure is known as an isobar.

23. The closer together the isobars, the stronger the wind speed. Hence widely-spaced ('slack' pressure gradient) isobars indicate light wind speeds. Close isobars ('tightly packed', steep pressure gradient) indicate high wind speeds.

24. Coriolis force deflects moving air to the right in the northern hemisphere.

25. According to Buy Ballot's law, in the northern hemisphere, if you stand with your back to the wind, the area of low pressure is to your left.

26. A change of wind direction when the wind direction changes clockwise (increasing), is a veering of wind direction.

27. From 2000ft to a land surface, the wind direction backs (by, say, 30°) and decreases in strength by around 50%. Thus, a 2000ft wind of 350/20 can be anticipated to become a surface wind of approximately 320/10.

28. The situation described (coastal airfield, cloudless summer day with light winds) is the classic set-up for a sea breeze. If the airfield is on the south coast, a sea breeze will blow from the south, possibly veering to south west later in the day.

M5 Clouds and Precipitation

29. Water can exist as a gas (water vapour); a liquid (eg cloud droplets, rain); or as a solid (eg ice, snow).

30. If air does not resist vertical motion, it can be said to be unstable. Flying conditions are likely to be 'bumpy', and cloud is likely to be cumulus type and have significant vertical extent.

31. In the lower atmosphere, any cloud in stable conditions are likely to be 'stratus' type.

32. The difference between temperature (14) and dewpoint (6) is 8 degrees. 8 x 400 = 3200ft; this is the anticipated base of any convective (eg cumulus) cloud.

33. Sheet-like low-level cloud, mostly uniform in appearance and possibly covering the whole sky, is likely to be Stratus-type cloud.

34. Thunderstorms are found within Cumulonimbus cloud (CB). Not all CB's contain a thunderstorm, but all thunderstorms are formed within a CB.

35. The meteorological term 'scattered' (SCT) indicates a cloud amount of 3 – 4 oktas, covering 3/8th to 4/8th of the sky.

36. The meteorological code 'DZ' used in aviation weather observations and forecasts denotes drizzle – precipitation in the form liquid water droplets, the droplet size being smaller than rain.

37. The code '+GR' indicates heavy (+) hail (GR). Out of interest, the abbreviation 'GR' is from the French word for hail – grêle.

M6 Visibility

38. The code 'BR' is used to indicate mist (from the French term for mist – brume).

39. Advection fog is most likely to form when a moist air mass moves over a colder surface, such as the sea or snow-covered ground. Advection fog tends to cover a large area and varies little by day or night.

40. Radiation fog is most likely to form overnight in Autumn (and to a lesser extent Spring) when the following conditions exist:
    - Little or no cloud at night and cooling of the land surface by radiation;
    - High relative humidity;
    - Light winds, ideally 2-8 knots.

41. HZ is the abbreviation for haze. In daylight condition, visibility in haze will tend to be worse when looking into sun, especially if the sun is low in the sky.

42. Heavy falling snow can quickly reduce visibility to a few hundred metres or even less.

M7 Air Masses

3. A Tropical Maritime air mass is predominantly warm and humid, and usually brings a stable air flow on reaching Europe. In winter, a tropical maritime air mass tends to bring dull weather with low stratus cloud, poor visibility and possibly advection fog – especially along south-west facing coasts.

4. A polar maritime air mass originates over the Atlantic near the Canadian Arctic or Greenland. On its passage southwards over the Atlantic, this air mass is heated from below by the warmer seas, so the moisture content of the air mass increases and it tends to become unstable. Arriving in Europe, polar maritime air brings convective (cumulus-type) clouds which are often well developed and can produce showers or even thunderstorms.

M8 Low Pressure Systems

5. A front is the boundary between two different air masses.

6. In general, depressions form under the jetstream and follow a track similar to that of the Jetstream.

7. A front bringing warm air to replace cooler air is always a warm front.

8. The passage of a cold front is often marked by a dramatic change in weather. The cloud and rain often disperse quickly, leaving drier and much clearer conditions behind the cold front. The temperature and dew-point fall but the pressure rises. The surface wind often veers sharply and becomes 'gusty' with the passage of a cold front, as a polar maritime air mass is introduced.

9. An elongated area of low pressure is called a 'trough'.

M9 High Pressure Systems

0. An anticyclone is an area of high pressure, where the air is generally descending and compressing.

1. The situation of temperature increasing with increasing height is that of a Temperature Inversion (the temperature lapse rate has been inverted).

2. Anticyclones usually bring light winds and (initially at least) clear skies. Over land in the Autumn, this situation may lead to the formation of radiation fog.

3. A ridge is an elongated are of high pressure.

4. A col (the area between two highs and two lows) has widely-spaced isobars; this leads to light winds. Just about all other weather that may occur in a col will be heavily influenced by the 'strongest' of the nearby weather systems.

M10 Hazardous Weather Conditions: Icing

5. Although airframe icing is a risk at any time an aircraft is flown in air temperatures below freezing (0°C) in or close to visible moisture, the greatest risk is considered to be generally when the air temperature is between +2°C and -20°C.

56. When an aircraft meets a supercooled water droplet, some of the droplet freezes on contact and forms a brittle layer of ice, usually on the leading edges of the wings and tail and also on aerials, struts, etc. This semi-opaque coating is called rime ice.

57. Clear ice is a sheet of almost transparent ice that forms on surfaces as a water droplet flows back from the leading edge and freezes.

58. Hazards associated with flight in icing conditions in a non-deiced aircraft include:
    - Increased weight
    - Increased drag
    - Change in aircraft Centre of Gravity
    - Possible ice blockage of engine intakes, pitot heads, static ports etc.
    - Possible moisture freezing inside pitot and static lines
    - Obscured windscreens
    - Freezing of flying controls
    - Increased stall airspeed
    - Loss of control

59. Rain ice is most likely to occur ahead of a warm front in the winter, in air temperatures below freezing.

60. Rain Ice is reported and forecast using the meteorological code FZRA (**F**ree**Z**ing **RA**in)

61. <u>ALL</u> frost (or ice or snow) must be removed from all aircraft surfaces before flight.

62. It is <u>NOT</u> necessary for the outside air temperature to be below freezing for carburettor to occur, as there is a large temperature drop within the carburettor itself. Carburettor icing may occur over a wide range of conditions, from +30°C to -10°C. In general high humidity is a greater risk than any precise air temperature.

63. Carburettor icing is most likely to occur at low power settings (such as those used during a descent).

M11 Hazardous Weather Conditions: Thunderstorms

64. The three key factors in thunderstorm development are:
    - An *unstable atmosphere*, with the instability reaching to the upper layers.
    - A supply of *moist air*.
    - Some form of *lifting action* to trigger-off the thunderstorm.

65. TCU is the aviation meteorological code for **T**owering **Cu**mulus

66. Thunderstorms can only occur within a Cumulonimbus (**CB**) cloud.

67. Thunderstorms are considered to have three key stages: Cumulus; Mature and Dissipating.

68. A heavy thunderstorm with rain has the meteorological code +TSRA.

69. Hazards to aircraft associated with thunderstorms include:
    - Heavy precipitation such as rain, snow and hail
    - Poor visibility is heavy precipitation
    - Turbulence
    - Windshear, downdraughts, microbursts
    - Strong winds

- Tornadoes (maybe visible as a funnel cloud or waterspout)
- Lightning
- Icing

0. The 'golden rule' to mitigate the hazards associated with a thunderstorm is avoidance – remaining a minimum of 10 nautical miles from the edge of the storm (not forgetting any overhanding anvil). Hazards such as gusts and turbulence may be experienced up to 20 nautical miles from a storm.

1. It is NOT safe to fly under a thunderstorm – even if you can see clear through to the other side – almost all of the thunderstorm hazards may occur without warning.

2. It is NOT safe to take-off with a thunderstorm over the departure airfield – hazards such as downdraughts, windshear and turbulence are a particular danger to an aircraft taking-off even if there is no lightning or precipitation occurring.

M12 Other Hazardous Weather Conditions

3. A general airflow can be significantly altered when it passes through a mountainous area. The wind direction may be changed to flow down a valley and 'funnelling' of the airflow through a valley or mountain pass may lead to an increased windspeed.

4. A katabatic wind may occur at night when a sloping surface cools; air in contact with the surface is also cooled, becomes more dense and flows down the slope. The resulting cold wind can be regular enough to have a local name in various parts of the world (for example the Bora in the Adriatic region of Italy, Slovenia, and Croatia).

5. It is NOT necessary for cloud to be present when mountain waves are occurring; when they do occur, lenticular clouds indicate that active mountain waves are present.

6. The area just below and downwind of a ridge or mountain summit is particularly dangerous in strong wind conditions, when rotor (severe turbulence) and downdraughts may be present.

7. A rain shadow is an area of higher cloudbase, better visibility, increased temperature and reduced precipitation than that generally occurring, often found on the downwind side of mountains or high ground. It is a practical illustration of the Föhn effect.

8. Windshear is a sudden change in wind velocity (direction and/or speed) over a short vertical or horizontal distance. Windshear may have an adverse effect on an aircraft's airspeed and flightpath.

9. Widespread gales are associated with a 'deep' depression (area of very low pressure), in particular if the pressure is continuing to drop, and at a rate of more than 24hPa in 24 hours.

M13 Meteorological Information

30. A cold front; the pressure at the centre of the associated depression is 959 hectopascals – L959

31. The brightest yellow/orange/red colours on a weather radar will, under normal protocols, signify areas of heavy or intense precipitation.

82. EMBD is an abbreviation of 'embedded'; usually referring to cumulonimbus cloud and indicating that CBs are expected to be embedded within cloud layers and so not readily seen or recognised.

83. A METAR (**Met**eorological **A**erodrome **R**eport) is an observation, a report of conditions actually occurring at a specified location at the specified time.

    A TAF (**T**erminal **A**erodrome **F**orecast) is a prediction, an estimate, of expected conditions likely to happen during a period in the future.

84. REFZRA = **RE**cent **F**ree**Z**ing **RA**in

85. 25043G58KT = surface wind from 250° true; wind speed 43 knots gusting to 58 knots

    3500 = prevailing visibility 3500 metres

    R27/0500 = Runway Visual Range along runway 27 of 500 metres

86. PROB40 = 40% probability

    TEMPO 2812/2816 = temporarily between 1200 and 1600 UTC

    0600 = 600 metres visibility

    +SHSN = Heavy **sh**owers of **sn**ow

    VV/// = vertical visibility indefinite ceiling (cannot be measured)

87. PROB30 = 30% probability

    TEMPO 2403/2406 = temporarily between 0300 and 0600

    0300 = 300 metres visibility

    -DZ FG = slight **d**ri**z**zle and **f**o**g**

    BKN001 = cloud broken at 100 feet above the airfield

88. The Aeronautical Information Publication (AIP) for a state will have a 'GEN' (General) section which will contain a description of 'Meteorological Services'; including the contact details for the organisation responsible for the providing aviation meteorological services in that state.

89. Automatic Terminal Information Service (ATIS) is a recorded radio broadcast of the observed weather and other aeronautical information for an airfield. A pilot would normally obtain ATIS information pre-departure or on approach on the designated radio frequency; although at some airfields the ATIS broadcast can also be obtained on a dedicated telephone number.

90. A transmission of recorded METAR reports for a group of airfields (usually within a specific geographical area) is a VOLMET broadcast.

M14 National Procedures

91. The designated source of official aviation meteorological information in the UK is the UK Met Office (www.metoffice.gov.uk).

92. At the box designated 50N 0230W the 5000ft wind is 060° True at 5 knots, the air temperature is +5°C.

93. It is impossible to answer this question. The date of flight is 31st May, but the forecast chart is dated 30th May – ie it is yesterday's weather chart.

M Appendix 1– Meteorology abbreviations and glossary

C	Altocumulus
D	Aerodrome
GL	Above ground level
MD	Amend or amended
MSL	Above mean sea level
S	Altostratus
TIS	Automatic Terminal Information Service
UTO	Automated METAR
ASE	Cloud base
CFG	Fog patches
ECMG	Becoming
KN	Broken (5 to 7 oktas)
L	Blowing
LW	Below
R	Mist
TN	Between
	Degrees Celsius
AT	Clear Air Turbulence
AVOK	Ceiling and visibility OK
B	Cumulonimbus
C	Cirrocumulus
I	Cirrus
IT	Near or over large towns
LD	Cloud
NL	Cancelled
OR	Correction
OT	At the coast
S	Cirrostratus
U	Cumulus
EG	Degrees
P	Dew point temperature
R	Low drifting (followed by DU, SA or SN)
S	Duststorm
U	Dust (in suspension)
UC	Dense upper cloud
Z	Drizzle
	East or easterly
NE	East-north-east or east-north-easterly
SE	East-south-east or east-south-easterly
MBD	Embedded
TA	Estimated time of arrival
TD	Estimated time of departure
FC	Funnel cloud
FCST	Forecast
FEW	Few (1 or 2 oktas)
FG	Fog
FL	Flight Level
FM	From
FPM	Feet Per Minute
FRQ	Frequent
FT	Feet
FU	Smoke
FZ	Freezing
FZUP	Freezing Unidentified Precipitation
G	Gust
GEN	Generally
GND	Ground
GR	Hail
GS	Small hail or snow pellets
H	High pressure centre
hPa	Hectopascals (= Millibar)
HVY	Heavy
HZ	Haze
IC	Ice Crystals (diamond dust)
ICE	Icing
INTSF	Intensifying
IR	Infrared
ISA	International Standard Atmosphere
ISOL	Isolated
JTST	Jet Stream
KM	Kilometres
KT	Knots
L	Low pressure centre
LAN	Inland (or over land)
LCA	Locally
LSQ	Line Squall
LV	Light & variable (relating to wind)
LYR	Layer(s), layered
M	Less than 0°C (temperature)
M	Metres
MAX	Maximum
MET	Meteorology
METAR	Aviation routine weather report
MI	Shallow (followed by FG)
MNM	Minimum

MOD	Moderate
MON	Above or covering mountains
MPS	Metres per second
MS	Minus
MSL	Mean sea level
MTW	Mountain waves
N	North or northerly
NC	No change
NCD	No Cloud Detected
NDV	No Directional Variation
NE	North-east or north-easterly
NIL	None
NM	Nautical Miles
NNE	North-north-east or north-north-easterly
NNW	North-north-west or north-north-westerly
NOSIG	No Significant Change expected
NS	Nimbostratus
NSC	No Significant Cloud
NSW	No Significant Weather
NW	North-west or north-westerly
OAT	Outside Air Temperature
OCNL	Occasional, occasionally
OVC	Overcast (8 oktas)
PL	Ice pellets
PO	Dust devils
PR	Partial
PROB30	30% Probability
PROB40	40% Probability
PS	Plus
QFE	Altimeter sub-scale setting to display height above a specified point on the service
QNH	Altimeter sub-scale setting to display altitude Above Mean Sea Level
RA	Rain
RASN	Rain and Snow (sleet in the UK)
RE	Recent
REUP	Recent Unidentified Precipitation
RMK	Remarks
RWY	Runway
S	South or southerly
SA	Sand (in suspension)
SC	Stratocumulus
SCT	Scattered (3 or 4 oktas)
SE	South-east or south-easterly
SEA	At sea
SEV	Severe
SFC	Surface
SG	Snow grains
SH	Shower (followed by RA, SN, PL, GR, GS or combinations of these)
SIG	Significant
SKC	Sky clear
SN	Snow
SNOCLO	Snow Closed
SPECI	Aviation special weather report
SQ	Squall
SS	Sandstorm
SSE	South-south-east or south-south-easterly
SSW	South-south-west or south-south-westerly
ST	Stratus
STNR	Stationary
SW	South-west or south-westerly
T	Temperature
TAF	Terminal aerodrome forecast
TCU	Towering cumulus
TEMPO	Temporary, temporarily
TL	Until
TOP	Cloud top
TROP	Tropopause
TS	Thunderstorm
TURB	Turbulence
UP	Unidentified Precipitation
UTC	Co-ordinated Universal Time
VAL	In valleys
VC	In vicinity of aerodrome (followed by FG, FC, SH, PO, BLDU, BLSA, BLSN)
VIS	Visibility, visible
VRB	Variable
VSP	Vertical speed
W	West or westerly
WDSPR	Widespread
WNW	West-north-west or west-north-westerly
WRNG	Warning
WS	Wind shear
WSPD	Wind speed
WSW	West-south-west or west-south-westerly
WX	Weather
Z	Co-ordinated Universal Time (UTC)

M Supplementary Study Material – Climatology

‘easonal Atmospheric Circulation Over Europe

.lthough Europe covers a very wide range of geography and conditions, from the /lediterranean in the south to the arctic circle in the north, from the western seaboard to astern continental plains, it is possible to make some generalisations about the climate and ow it effects the area as a whole. These generalisations have to be viewed as such; the /eather can, and does, change dramatically from day to day. Snowfall in the summer and pring-like days in the midst of winter are not unknown and even if neither lasts for more han a day it can be enough to change flying plans.

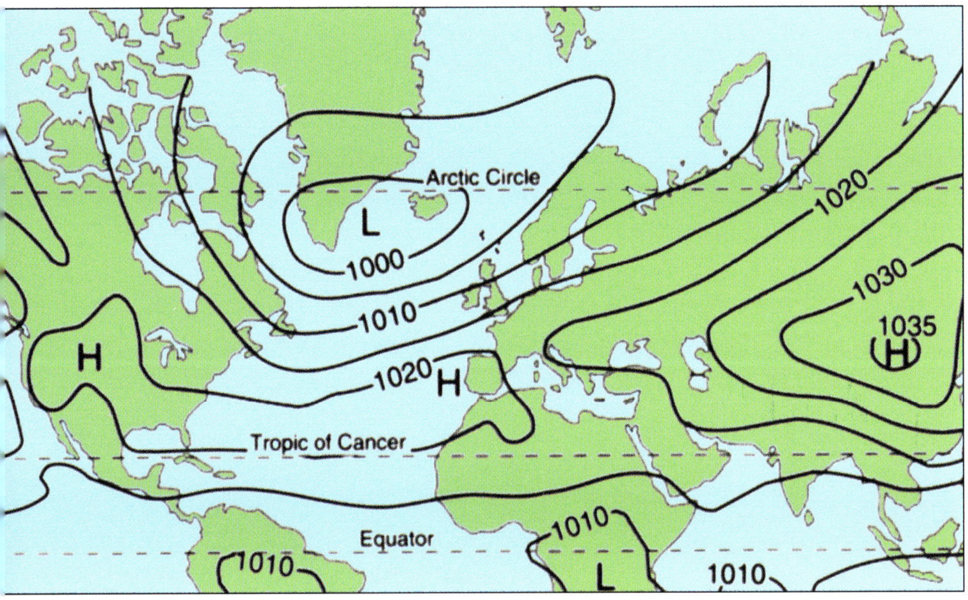

Figure App 1.1
The average pressure distribution over Europe during winter.

In the winter the predominate climate features are the low-pressure depressions that orm in the North Atlantic and ‘deepen’ as they sweep in from the west, typically bringing vith them cloud and precipitation and occasionally gales. These travelling depressions ollow a path largely dictated by the upper wind patterns and in particular the ‘jetstream’, vhich tends to lie further to the south in the winter than it does in the summer. The lower he pressure in a depression, the faster it moves and the quicker it deepens (ie the faster he pressure falls within the depression), the more active it will be and the worse the veather associated with it. On continental Europe the Alps mountain range can form a arrier to the movement of fronts, with cloud and precipitation persisting along and beyond he western side of their north/south axis. Between the passage of one depression and the rrival of the next there may be a short period of settled weather in a ridge of high pressure.

During the winter there is typically a large area of high pressure settled over continental astern Europe. At times this anticyclone may move over western Europe, ‘blocking’ the lepressions from the Atlantic. This anticyclone will typically bring clear skies initially, but lso plummeting temperatures in bitterly cold easterly and north-easterly winds. After a few lays low cloud is likely to form within this area and days or even weeks of gloomy weather ınder unbroken low cloud, possibly with drizzle, can follow.

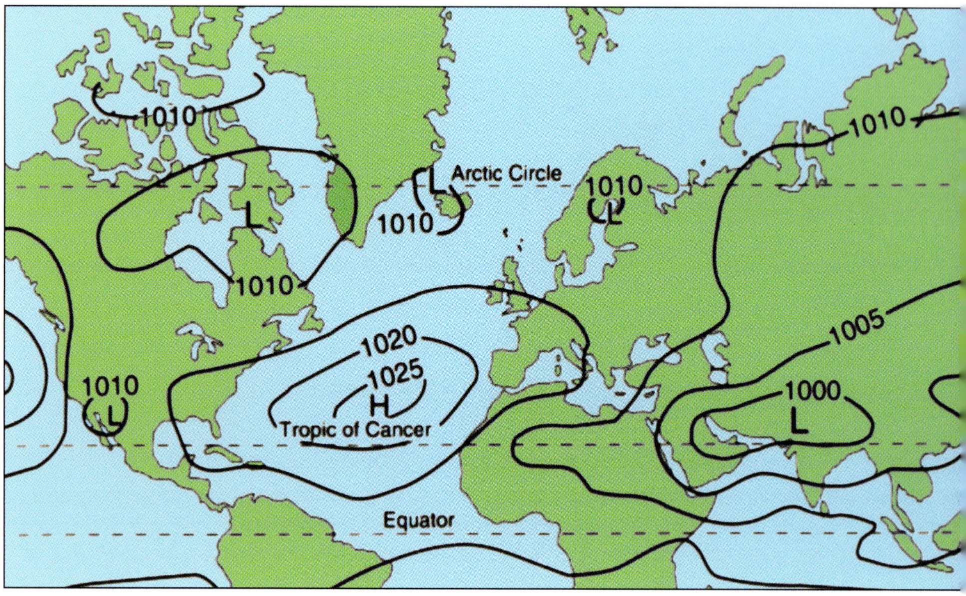

Figure App 1.2
The average pressure distribution over Europe during summer.

In the summer the European climate is likely to be dominated by an area of high pressure pushing up from the mid-Atlantic, often known as the 'Azores high'. The travelling depressions that do still form in the North Atlantic tend to be less active and track further to the north than in the winter. If high pressure does extend up from the Azores area it generally brings fine and settled weather and, at the height of summer, it can bring a 'heatwave'. Although visibility is usually good at first, haze and smoke can build up as an anticyclone becomes established. Visibility, particularly under a marked inversion, may well worsen until the high pressure breaks down or moves away.

Local Seasonal Weather and Winds

Localised weather and winds are almost impossible to summarise, it may take years of experience of a particular location to recognise and be able to predict them. This is why sometimes an experienced pilot who knows the local area well can make a more accurate assessment of local weather trends than a remotely-based forecaster. As always, the best advice in this respect is to always err on the side of caution Locally-based pilots, ATSU personnel or forecasters may be able to give sound advice about local conditions.

Some localised weather and wind phenomena are relatively well-known, such as the 'Mistral' valley wind that brings cold and turbulent northerly winds down the Rhone valley, or the 'Bora' katabatic wind in the Adriatic. Less well-known localised conditions occur throughout Europe. In northern Europe, west and south-facing coasts are particularly vulnerable to low cloud and precipitation coming in from the Atlantic on the prevailing westerly winds. South-westerly winds bringing tropical maritime air often mean fog or low cloud, especially in the spring and early summer, arriving almost without warning as the moist airflow arrives on the landmass. Conversely, land areas to the east of mountains and high ground tend to have drier and brighter conditions, and in general there is less average precipitation as one moves eastwards across Europe.

The Mediterranean has almost a 'mini-climate' of its own, often quite different to conditions relatively close by. The winter weather can be surprisingly wet and unsettled, especially along the northerly shores. The summer weather is far more settled, dry and hot. However, when thunderstorms do occur, they can be particularly violent. It is also notable how regularly a dry spell in the northern Mediterranean breaks down into storms, if only temporarily, in around mid-August.

The best practical advice is to collect as much information as possible about both what the weather is actually doing and what it is forecast to do, and not to assume that weather climate and trends at the home airfield are applicable further afield. The knowledge of a local pilot or local weather forecaster can be invaluable in making safe flying decisions in an unfamiliar area.

A Final Word about Aviation and Weather

All the meteorological knowledge and aviation wisdom in the world is of no use at all to a pilot who is not prepared to apply it. Every year, serious accidents involving General Aviation aircraft are caused, at least in part, by weather factors. So, if at some stage a real doubt creeps into your mind about the chances of safely completing a flight, *right there and then is a good time to start reviewing your alternatives,* whilst they are still open to you.

Sometimes simply deciding not to commence a flight because the actual or forecast weather is unsuitable is the best decision a pilot can make. If in-flight conditions become markedly worse than forecast – and if you are stretching the limits of your licence, abilities and the definition of VFR to carry on – *now* is the time to turn back or divert. All experienced VFR pilots occasionally have to cancel or abort a flight, or (more rarely) make an in-flight diversion. Deciding to do so is a sign of good airmanship and sound judgement, and a good application of the principles of Threat and Error Management (TEM). Weather forecasts will occasionally be wrong – nature feels no obligation to conform with a forecast, especially at a specific location; circumstances will sometimes conspire against you. That's life: deal with it. Take the safest option and value the experience you gain in the process.

There is always a degree of pressure to get back to your home airfield, or to complete the planned flight against the odds. Be aware that this pressure can very easily cloud your judgement and impair your decision-making. The age-old pilot diseases of 'press-on-itis' and 'get-home-itis' have been responsible for destroying more perfectly serviceable aircraft (and their occupants) than anyone can count. Don't let it happen to you. Human nature is such that a more experienced pilot – or some other strong personality – can exert forceful pressure, whether consciously or not. The result can be that you either commence a flight against your better judgement, or you continue into conditions you are not comfortable with. However, as Pilot In Command *you must always judge for yourself the safest course of action.* Remember, VMC criteria are an absolute limit, not a target; and in many circumstances safe VFR flight becomes difficult to achieve before the VMC limits are reached.

It is the case that dealing with bad weather is all about pilot judgement and virtually nothing to do with flying skill. No amount of piloting ability will make fog less dense, reveal a mountain hidden by cloud, or make the inside of a thunderstorm less turbulent. The root cause of 'weather' accidents is nearly always lack of judgement: either by ignoring or failing to account for forecast bad weather before take-off, or by continuing the flight into poor weather, rather than diverting or turning back. Taking on the weather, maybe 'having a go', is merely inviting the inevitable conclusion to a one-sided contest. After-all, weather is just a process of nature – and nature always wins, sooner or later. The losers are those pilots who become just another 'weather accident' statistic.

More often than not, weather-related accidents are shown to be totally avoidable. Flying is not cheap, and there is little point in spending money to frighten yourself and your passengers or to put yourself or others at risk. Use your meteorological knowledge to make good decisions and build sound pilot judgement, and above all **enjoy** your flying.

M Index

Notes